Polar Covalence

Academic Press Rapid Manuscript Reproduction

Polar Covalence

R. T. Sanderson

Emeritus Professor of Chemistry
Arizona State University
Tempe, Arizona

1983

ACADEMIC PRESS

A Subsidiary of Harcourt Brace Jovanovich, Publishers

New York London

Paris San Diego San Francisco São Paulo Sydney Tokyo Toronto

COPYRIGHT © 1983, BY ACADEMIC PRESS, INC.
ALL RIGHTS RESERVED.
NO PART OF THIS PUBLICATION MAY BE REPRODUCED OR
TRANSMITTED IN ANY FORM OR BY ANY MEANS, ELECTRONIC
OR MECHANICAL, INCLUDING PHOTOCOPY, RECORDING, OR ANY
INFORMATION STORAGE AND RETRIEVAL SYSTEM, WITHOUT
PERMISSION IN WRITING FROM THE PUBLISHER.

ACADEMIC PRESS, INC.
111 Fifth Avenue, New York, New York 10003

United Kingdom Edition published by
ACADEMIC PRESS, INC. (LONDON) LTD.
24/28 Oval Road, London NW1 7DX

Library of Congress Cataloging in Publication Data

Sanderson, R. T. (Robert Thomas), Date
   Polar covalence.

   Includes bibliographical references and index.
   1. Chemical bonds. I. Title.
QD461.S344 1983   541'.224   82-24432
ISBN 0-12-618080-6

PRINTED IN THE UNITED STATES OF AMERICA

83 84 85 86   9 8 7 6 5 4 3 2 1

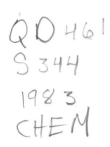
Contents

Sources of Most Useful Data

Preface

Here is an up-to-date account of a uniquely successful approach to understanding chemistry from a knowledge of atomic structure and the properties that result from this structure. It is simple enough to be understood by beginning students and practical enough to have value for all professional chemists. Its validity has been established beyond any reasonable doubt. In all these respects it stands alone in modern theoretical chemistry. To be extremely useful, it requires only to be used. Certainly there are imperfections, but they should serve as challenges to researchers seeking improvements, not as excuses for rejection.

Except for six M8 elements, helium, neon, argon, krypton, xenon, and radon, all substances on this earth exist as atoms joined together. The properties of each substance are the consequence of the qualities inherent in the structure of the individual component atoms and of the manner in which these atoms are bound together. All chemical reactions involve the rearrangement of atoms, breaking the bonds in reactant elements or compounds, and, usually, forming new bonds. The direction of chemical reaction tends toward greater entropy and greater average bond strength. At ordinary temperatures the tendency toward greater bond strength usually predominates, unless the increase is quite small and the entropy decreases quite substantially in that direction. We may study the individual reactants, examine the individual products, establish optimum conditions for the reaction to proceed, and learn what we can about the kinetics and thermodynamics, but we will never really achieve the illusion of understanding chemistry until we know *why* certain bonds are stronger than others. Only then can we appreciate the origin of the urge or resistance of the combined atoms to rearrange. Therefore, nothing seems more fundamental to understanding of chemical bonds.

Fortunately, such an understanding has not been essential to marvelous applications of chemical science, which have developed from an enormous body of experimental work. We know from experience, of course, that there is a great re-assuring consistency in chemical phenomena, that if the reactants are identical and the experimental conditions carefully set, the same results will be observed over and over again. But even though immensely useful and gratifying results can be obtained without our understanding exactly why, the question is always there. Whereas most of us are reasonably content to satisfy our curiosity as to what will happen with these new materials or under these new conditions, a substantial number of us cannot rest until we have done our best to understand *why* it happens. Theoretical explanations and practical results are mutually supportive. Practical results reinforce theories or suggest improvements, and theories suggest new experiments that provide practical results.

It seems strange but true that even scientists must continually be reminded that what is not known is not predictable. The principal goal of theorists seems to be to learn to understand well enough to predict the results of experiments not yet performed. Theories are commonly tested on the basis of their ability to predict. Yet in fact, if some unknown is truly predictable, this is almost the same as if we had known it all along. The *truly unknown is unpredictable.* The creative scientist is not the one who verifies by experiment what was a foregone conclusion, or performs routine measurements on a substance not previously tested. He is the one who can imagine novel exploration into the unknown and then control his preconceptions sufficiently to recognize the full significance of whatever results occur.

The point is, if what we may discover really *cannot be predicted,* then surely the best pathway to this discovery *cannot be predetermined.* Until the goal has been attained, we ought never to abandon any imaginable means of reaching it. However, it is human nature to seek success, and if any particular route appears more favorable, this is where the effort will be concentrated. Other routes will tend to remain relatively unexplored. Nearly sixty years ago, theoretical chemists began to approach the problems of chemistry through quantum mechanics, and over the years they seem to have developed a firm conviction that only by this means can the truth be discovered. Other possibilities, including the approach described in this book, seem to have been ignored.

This attitude might deserve sympathy if, in the course of those sixty years, a thorough understanding of chemical reaction through quantum mechanical calculations of bond energies and bond lengths had been developed. It has not. Moreover, such success may possibly never be achieved. First, the mathematics of interrelationships among many bodies is immensely complex and requires elaborate approximations for results of practical value, eliminating the possibility of easy understanding. Second, the total energy of a molecule or the atoms that compose it is usually extremely great compared with the energy of bond formation. Bond energy represents the difference between the total energy of the separate atoms and that of the molecule, a relatively small difference between very large values. It would require calculation of the large values with probably unattainable accuracy in order to obtain useful bond energy evaluation. For such reasons, it has long appeared to me imperative to examine other possible approaches, *not to challenge or deny the truths of quantum mechanics,* but merely to *fill a void that needs filling.* I hope this is a satisfying answer to the reasonable question: Why does anyone interested in explaining chemistry stray from the universally approved path? *You can't get there from here unless you do.*

This book provides a practical alternative in which oversimplification and lack of rigor are, I believe, amply atoned for by perceptive intuition, ease of understanding, and remarkably accurate quantitative results. It has long seemed to me that there is a duality to the reality of our universe, in which infinite complexities are interwoven with marvelous simplicities. Surely there is noble purpose in attempting to unravel the complexities, but also it seems foolish to ignore the simplicities that can contribute so much to a practical understanding. One must sometimes choose between a simple-minded approach that works and a sophisticated approach that does not. This book brings up to date the results of continuing studies since my two editions of "Chemical Bonds and Bond Energy." It is my hope that I can at last persuade my fellow chemists of the practical value and inherent reliability of this approach.

Continuing research has established the theory of polar covalence on an even firmer basis. The following changes and additions will be noted:

(1) Based on a more thorough correlation with fundamental atomic properties, the electronegativities of the elements have been revised, along with the nonpolar covalent radii and the homonuclear single covalent bond energies. As a concession to common practice and in the hope of encouraging more willing acceptance, the scale has been changed to resemble that originated by Pauling, in that fluorine, the most electronegative of all the elements, has the value 4.000.

(2) The new values have been used to recalculate the bond energies in more than one thousand molecules, the average agreement between calculated and experimental atomization energies being within 1 kcal/mole of bonds.

(3) The Lone Pair Bond Weakening Effect has been studied further, and although not yet thoroughly explained, the limits of applicability have been more clearly delineated.

(4) The nature of multiple bonds has been studied and the effects of disparity in atomic radius recognized and evaluated.

(5) Additional attention has been paid to the bonding in nonmolecular solids and in molecular addition compounds and organometallic compounds.

(6) Application to organic compounds has been modified in recognition of two minor but significant factors that influence the bond strength in hydrocarbon groups: bond strengthening by chain branching and bond weakening by repulsions between nearby branches. Atomization energies of more than 750 organic compounds of all common functional types have been calculated and found to be in very satisfactory agreement with experimental data.

(7) Based on improvements in the calculation of contributing bond energies in organic molecules and on improved data on heats of formation of common radicals, evaluation of reorganizational energies of radicals has been thoroughly

revised and applied to an improved understanding of bond strength as indicated by bond dissociation energies.

(8) The concept of radical reorganization energies has also been applied to gaseous inorganic molecules, including those exhibiting subnormal valences. This leads to a better understanding of the nature of bonding in many inorganic fragments such as the subhalides.

(9) Applications of these concepts to the periodicity of inorganic compounds are reviewed and summarized.

(10) Last but not least, numerous areas of ignorance are exposed to challenge others in their future researches.

The basic principle of this work, the principle of electronegativity equalization, has now been confirmed through quantum mechanics by Robert G. Parr and his associates, to whom I am very grateful. This was about twenty-seven years after initial publication of the principle in 1951. Being at this time of writing in my 70th year, it seems unlikely that I shall be available to applaud confirmation of these further developments twenty-seven years hence. However, I really do not believe such confirmation, although highly desirable, is truly needed for practical application now. The theory of polar covalence is so strongly supported by experimental evidence that lingering but unsubstantiated doubts of its validity can no longer justly delay its practical use. Surely it can provide a greater depth of understanding to ordinary chemists than they can possibly have without it.

This book is dedicated to my late wife, Bernice, and to my present wife, Jean, two splendid women who have given me strength and encouragement to keep trying under difficult but challenging circumstances. I am grateful also to the many teachers and students who have seemed appreciative of my efforts to explain chemistry better, and to the generations of chemists whose years of work have provided most of the indispensable raw material to which I have tried to give organization and meaning.

R. T. Sanderson

4725 Player Drive
Fort Collins, Colorado

Chapter 1

CHEMICAL BONDING: A GENERAL DISCUSSION

Of the several types of force recognized in nature, only those between electrical charges appear significant in ordinary chemistry, which is concerned with rearrangements of atoms and does not involve nuclear change. Chemical bonds of course involve attractive forces, which means between unlike charges. All atoms are normally electrically neutral. Nevertheless, in order for unlike charges to be responsible for joining atoms together, there must be some opportunity for the positive nucleus of one atom to attract one or more electrons of another atom.

The conditions that provide such an opportunity are suggested by the existence of six elements whose atoms do not and the rest which do. These six are the M8 (major group eight) elements, helium, neon, argon, krypton, xenon, and radon. The last five are the only elements whose atoms have 8 outermost electrons, completely filling the s and p orbitals (helium of course having only 2 outermost electrons completely filling the s orbital in the first principal quantum level). All other elements have fewer than 8 outermost electrons (and in hydrogen, fewer than 2). The quantum theory of atomic structure describes the fact that electrons surrounding a nucleus must occupy certain specified regions called orbitals, within which they can sense a net attraction for the nucleus despite the repulsions of the other electrons. The most important outermost orbitals are four in number, one s and three p, and can accommodate 2 electrons each. Wherever the number of outermost electrons is less than 8, the presence of vacancies is implied, the number of vacancies being 8-n where n is the number of outermost electrons.

A **vacancy** is a region having no electron but within which an electron would sense a net attraction for the atomic nucleus. Such vacancies are extremely important in chemistry, for it is their **existence** in atoms which permits the atoms to form chemical bonds to other atoms, and their **absence** which makes atoms inert. Only fluorine, the most electronegative of the chemical elements, has atoms capable of attracting electrons so strongly that they even activate the atoms of the heavier M8 elements which normally have

1

no stable outer vacancies. Whenever an electron of one atom enters a vacancy of another atom, the essential condition for bonding is created, for the same electron is now under the influence of both nuclei. **All chemical bonding** is essentially of this nature: **the same electrons are simultaneously attracted to two or more atomic nuclei.** However, there are numerous variations of this theme. Some of these can be described simply and quantitatively and some still defy our best efforts.

NONPOLAR COVALENCE

The simplest example of a chemical bond is that which holds together two atoms of hydrogen in an H_2 molecule. Each hydrogen atom has one electron, accommodated in an orbital capable of holding two electrons. Thus each atom has also one vacancy. This vacancy, remember, is a region within which an electron from another atom would sense a net attraction for the nucleus, despite the repulsion by the other electron. When two such atoms come together, the electron of each can be accommodated within the vacancy of the other. The two electrons, called bonding electrons, are now held by both nuclei. It is this attraction of both nuclei for the same two electrons which constitutes the chemical bond. The two hydrogen atoms are held together very securely. It will require, at ordinary temperatures, 104.2 kcal (436.0 kJ) per mole of H_2 molecules to restore the state consisting of free hydrogen atoms. This is called the bond energy, and since the two atoms are alike, it is the **homonuclear** bond energy. The two nuclei are separated by an equilibrium distance of 74 picometers, which is called the bond length. Bond length and bond energy are closely interrelated, since the bond energy is essentially a coulombic energy obeying the laws of electrostatics. According to Coulomb's law, the force of attraction is proportional to the product of the opposite charges and inversely proportional to the square of the distance between their centers. The Coulomb energy is similarly proportional to the product of the charges and inversely proportional to the distance between charge centers. Therefore shorter bonds tend to be stronger, since the electrostatic attractions operate at smaller charge separations.

The bond in hydrogen is called a covalent bond. Because

the two atoms are exactly alike, they must attract the bonding electrons exactly equally. The bond is therefore called **nonpolar covalent,** meaning that the bonding electrons are shared **evenly** between the two atoms. We may consider the H_2 molecule as containing the prototype of the covalent bond. From it we deduce that the requirements for forming such a bond are first, that **each atom must provide one outermost electron to be shared with the other atom,** and second, that **each atom must provide one outermost vacancy capable of stably accommodating an electron from the other atom.** Already it becomes apparent why the M8 elements, helium, neon, argon, krypton, xenon, and radon, normally form no covalent bonds: **their atoms possess no outer vacancies capable of accommodating electrons from another atom where they can sense a net attractive force.** All the stable regions are fully occupied.

Ever since the discovery of these elements, and the recognition of their electronic configurations, it has been tempting to suggest that there is a special magic to having 8 outermost electrons (or 2 in the case of helium), a special stability associated with the octet. From this, it has in the past been supposed that atoms of the active elements tend to react in the direction of acquiring this special stability, as though their motivation in joining to other atoms were their inherent wish to become like an inert element in electronic structure. For example, an atom of sodium has 11 electrons, one beyond the octet. If it could lose this outermost electron it would become like neon, atomic number 10, with an outer octet of electrons. Therefore, it was argued, sodium tends to lose an electron to form sodium ion, Na^+. Similarly, an atom of fluorine has only 9 electrons, one short of the outer octet, and if it could gain one electron, which it certainly tends to do, it too would resemble the very stable neon atom. To this day some authors and other teachers still cannot resist the temptation to present chemical reaction in this manner, which seems so beautifully simple and logical.

Unfortunately, this argument is fallacious, as can easily be shown. The fallacy lies in the fact that the special stability of the octet of outer electrons depends on a particular number of positive charges at the nucleus. Neon is unreactive not merely because it has 8 outermost electrons but also because it has **exactly the right nuclear charge,**

+10, to hold these 8 electrons tightly without activating the next higher orbitals to an ability to attract electrons. If the outer octet is left intact, but one proton is removed from the nucleus of neon, the atom is converted to a fluoride ion. Unlike the neon atom and despite its possession of 10 electrons, the fluoride ion is capable of hundreds of possible chemical combinations. If an extra proton is added to the neon nucleus, without altering the outer "stable" number of 10 electrons, the atom becomes a sodium ion, again capable of hundreds of possible chemical combinations, or of gaining an electron to become sodium atom. With a nuclear charge of +11, the outer octet is no longer able to cover up the nucleus effectively, and the next higher outer orbital, 3s, becomes able to hold an electron. With a nuclear charge of only +9, the outer octet is no longer so stable and an electron can be lost to a positively charged electrode or partially lost to any cation. The magical number of 10 electrons represents "stability" **only** when the nuclear charge is also 10.

Why, then, do sodium atoms tend to lose one electron and fluorine atoms tend to gain one electron? Sodium tends to lose the outermost electron simply because nearly every other kind of atom can attract electrons more strongly. It never loses more than one electron because when it has lost one, the next electrons are contained in an octet which is held even more tightly than in neon, having a nuclear charge of +11 instead of only +10. Fluorine atoms tend to gain one electron because they have one outermost vacancy capable of accommodating it, within which a strong net positive charge can be sensed. They tend never to gain more than one electron because, with a nuclear charge of only +9, they are even less able than a neon atom to attract electrons to an outer orbital.

In general, atoms stop at the removal of just the outermost electrons because no chemical agent is capable of removing any of the tightly held underlying electrons. Atoms stop at the gaining of electrons when their outermost vacancies have in effect been occupied. For many, but by no means all, of the major group elements these limits do correspond to the stable octets of the inert M8 elements, but **not** because the active atoms have achieved the magical stability of the M8 atoms.

Having observed the minimum requirements for formation of a covalent bond, we can then understand how an atom might well form one covalent bond for each outermost halffilled orbital it can provide. This number of bonds is called the **valence** of the atom. Hydrogen clearly has a valence of one, with no chance of forming more than one bond per atom using normal covalence. Its well known ability to serve as a bridge between two other atoms involves deviations from the usual idea of covalence, which will be discussed later on. We can predict that atoms having but one outermost electron can form but one covalent bond, those having two outermost electrons can form two covalent bonds, up to four bonds by those atoms having four outermost electrons. At this point, the number of outermost vacancies is also four, so that a valence of four is expected. With five outermost electrons, however, there are only three vacancies left in the octet, so the valence is limited to three, and with six outer electrons the valence is two, and with seven outer electrons, the valence is one. These limitations may be expanded under certain conditions when outer vacant orbitals beyond the octet may become activated. They may be diminished when two outer electrons remain paired in the same orbital and thus unavailable for normal covalence.

In general we may expect each atom to form as many covalent bonds as it can. This largely determines the composition of molecules and compounds as well as the nature of the free elements. If we know that one atom has three outermost half-filled orbitals and another atom has but two, we can predict that any compound of the two will have three atoms of the latter combined for every two of the former, so that each element will be forming the same total number of bonds.

In order for a covalent bond to be effective, the bonding electrons must clearly spend most of their time somewhere within the region between the two nuclei. In this sense the bonding electrons are restricted--**localized**--within the internuclear region. This is a characteristic of covalence, that the bonding electrons are localized. But a characteristic of all electrons is that they repel one another. They will never remain localized if they have any option that will allow them to move, on the average, farther apart. And this brings us to a consideration of what happens when

formation of all possible normal covalent bonds leaves one
or more outermost orbitals still vacant.

METALLIC BONDING

When the number of outermost vacancies exceeds
the number of outermost electrons, then formation of the
maximum possible number of outer half-filled orbitals and
therefore bonds will still not make use of all the outer vacancies. In such a situation any bonding electrons have the
option to expand into the extra vacancies, thus becoming
delocalized. When the bonding electrons are delocalized,
they are no longer restricted to specific regions between
specific nuclei, but can spread out to minimize repulsions.
This changes the character of the bonding remarkably, for
it eliminates the restrictions on the number of bonds the
atoms can form. In most cases the atoms become as closely
packed as physically possible for like spheres. Indeed, the
concept of individual bonds is replaced by a concept of many
close-packed atoms held together by the mutual attraction
exhibited by their nuclei for the valence electrons which
are delocalized among them.

Since the physical properties of any substance depend
largely on the manner in which its atoms are interconnected,
the delocalized electron state corresponds to dramatic differences in the physical properties. The substance tends to
allow easy flow of electrons, since the bonding electrons
are no longer tied down to specific internuclear regions.
Easy flow of electrons corresponds to good conductivity
both of heat and electricity. If the electrons were restricted
to specific covalent bonds, there could be no appreciable
electron flow and the substance would be insulating both
thermally and electrically. If the electrons were restricted
to specific bonds, there could be no appreciable rearrangement
of the atoms without breaking some of the bonds, and therefore the substance would tend to be hard and brittle. But
with delocalized bonding, one layer of atoms can slide over
another layer without appreciable disruption of the bonding,
so the substance tends to be ductile and malleable. The
state of the combined atoms in which the bonding electrons
are delocalized is called the **metallic state,** and the bonding
is called **metallic bonding.** It represents a condition of the

atoms which is more stable than it could be if only the obvious localized covalent bonds could be formed.

Consider, for example, the vapor of sodium, which consists largely of separate atoms. Each sodium atom has one outermost half-filled orbital, which entitles it to form a single covalent bond. As the vapor is cooled toward its condensation point, a small percentage of the atoms do combine to form diatomic molecules, Na_2, which have a dissociation energy of about 16.4 kcal (68.6 kJ) per mole. As further cooling occurs, these diatomic molecules do not remain as such but condense further to a liquid, and then a solid in which each interior atom is at the center of a cube of other atoms, and 15 per cent farther away from six more neighbors at the centers of the six surrounding cubes. This is the familiar body-centered cubic structure exhibited by a number of metals. Atomization of this metal requires about 25.9 kcal (108.4 kJ) per mole, whereas atomization of the Na_2 molecules requires only half of 16.4 or 8.2 kcal (34.3 kJ) per mole of atoms. Thus the metallic crystal is more than three times as stable as the diatomic molecules. Delocalization of the bonding electrons provides a distinct advantage over the nonpolar covalent possibilities of these atoms. In general we may regard the metallic state and metallic bonding as the preferred condition of like atoms, if it is possible for them. In this sense, nonpolar covalence occurs only when there is no alternative opportunity for the bonding electrons to spread out, escaping the confinement of the internuclear region.

On this basis it is easy to understand why all of the transitional and inner transitional elements are metals. Not one has more than 2 outermost electrons where 8 are possible, so each has plenty of vacancies into which bonding electrons can expand and become delocalized. It is also easy to understand why the elements of the first three major groups should also be metallic, since the outer vacancies in their atoms always outnumber the outer electrons. Boron is exceptional, presumably because its small size involves holding the electrons too tightly for normal metallic bonding. It commonly occurs in spherical clusters of 12 atoms each, in which each atom has five neighbors within its own sphere and is attach ed to one atom of another sphere. Thus boron represents a kind of compromise. If its atoms exhibited normal covalence, each atom would be surrounded by three other atoms at the corners of an equilateral triangle and the element would exist as parallel sheets of atoms of indefinite extent. If the element were truly metallic, it would have at least eight closest neighbors instead of six, and probably 12 as is typical of many metals which exhibit the closest packing possible for spheres of like size, and completely delocalized bonding electrons. But boron is the sole exception in the first three major groups.

Somewhat more difficult to explain is the existence of the metallic state in the heavier elements of groups in which the atoms have their outer four orbitals at least half full. For example, in Group M4, tin exhibits two forms, a grey nonmetallic form stable below about 13°C, which exhibits normal tetracovalence in the manner of diamond, and a white, metallic form stable above that temperature. There is some evidence suggesting that in the metallic state, the tin is actually divalent, only two of its outer four electrons being involved in the bonding. If two electrons remain paired, this leaves a vacant orbital which would allow delocalization of the bonding electrons in the metallic form. For lead no nonmetallic form has been observed, it being characteristically metallic. The greater tendency for two outermost electrons to remain paired is probably responsible here. For the M5 metals, antimony and bismuth, explanation is not so simple. It might be the consequence of greater availability of outermost d orbitals in these elements, or perhaps by double pairing of outer electrons these elements could be in effect monovalent metals.

POLAR COVALENCE AND "IONIC" BONDING

These characteristics of nonpolar covalent and metallic bonding account for the physical states of the chemical elements in a generally satisfactory way. Complications arise when the atoms are not alike. When they are not alike, they exhibit different degrees of attraction for bonding electrons. Thus when they form covalent bonds, the bonding electrons are not evenly shared between the two unlike atoms, but spend more than half time more closely associated with that atom which originally attracted them more. This imparts a partial negative charge on that atom, leaving the other atom with a partial positive charge. The bond is now called **polar covalent.** As will be seen, it is stronger than it would be if it were nonpolar. Since the most important tendency at ordinary temperatures is for atoms to rearrange in the direction of forming stronger bonds, this means that **chemical reaction tends to occur in the direction of greater bond polarity.** This probably explains why most of the matter on earth consists of compounds rather than pure elements.

Normally atoms tend to form as many covalent bonds as they supply half-filled orbitals. When they have combined to the full extent of this capability, a stable molecule or nonmolecular solid results. However, when the atoms are of different elements, maximum covalence does not necessarily terminate the possibility of bond formation. This is usually because full utilization of the bonding electrons of an atom of a metallic element leaves unused vacant orbitals still capable of attracting electrons. Similarly, full use of bonding electrons in a nonmetallic element always, except for M4 elements, leaves at least one lone pair of electrons that is not involved in the bonding. We shall consider later in detail the factors that determine the relative attraction for bonding electrons exhibited by any given element. For the moment it is sufficient to point out that this attraction is considerably lower for metal atoms than it is for nonmetal atoms. Therefore in any combination of the two it is the latter which acquires partial negative charge, leaving the metal with a partial positive charge.

The significance of a partial positive charge on any atom which has unused outer orbitals is that it enhances the attraction for the nucleus which can be sensed by any electron from another atom that enters such an outer vacant orbital. In other words, the combination of properties makes the atom a potential acceptor of electrons. On the other hand, a partial negative charge on any combined atom which also has a lone pair of electrons in its outer shell makes this atom a potential provider of the lone pair to occupy the vacant orbital of the acceptor. If, therefore, any neutral molecule which still has an outer vacant orbital on one of its atoms together with a partial positive charge, encounters another molecule which has an outer lone pair on a negatively charged atom, further combination can occur. Two unlike molecules can add together to form a molecular addition compound. The bond is called **coordinate covalent.**

Another possibility is open when acceptor and donor are part of a single molecule, a compound of metal with nonmetal, which has vacant orbitals left on the positively charged metal atom and lone pair electrons not involved in the bonding on the negatively charged nonmetal atom. For example, when sodium unites with fluorine, the simplest molecule that might form is NaF, in which the sodium still has three vacant orbitals and the fluorine now has three lone pairs of electrons. Other like molecules will tend to orient themselves so that each sodium is surrounded by fluorine and each fluorine atom by sodium atoms. Thus is formed a nonmolecular solid in which the identity of the original molecules has disappeared, each atom having six of the other kind of atom surrounding it at equal distances.

Where the partial charges on the combined atoms are high, these compounds are conventionally thought of as consisting of ions, as though the nonmetal atom had complete control of the bonding electrons, giving it negative charge, and leaving the metal atom as cation. The bonding is then called "ionic" and ascribed purely to electrostatic forces among the oppositely charged ions. Although the major part of the force holding the atoms together in such solids is ionic, it is doubtful if complete electron transfer occurs, there being also a significant contribution of covalent energy. The bonding in solids can in some instances be treated in a fairly simple manner and quantitatively, but the degree of coordination is variable and not yet predictable.

SUMMARY

Most of the chemical substances which qualify as compounds, by virtue of their being sufficiently stable to allow some of their properties to be measured experimentally, are held together by bonding of the types described. In the free elements, if there are sufficient vacancies in the outermost part of the atom for the bonding electrons to spread out and become delocalized, metallic bonding is observed. If there are not such vacancies, then the bonding electrons must remain localized essentially within the internuclear regions and the bonds are nonpolar covalent, leading to nonmetals either as molecules or nonmolecular solids. In substances containing more than one kind of atom, the bonds are polar covalent. This bonding may be supplemented by coordination to make fuller use of any available orbitals. **In all bonding, the fundamental force appears to be the result of more than one nucleus interacting attractively with the same electrons.**

In our search for a quantitative understanding of chemical bonds, we must seek the simplest concepts that are practically workable. It is not yet possible to account for the energies of homonuclear covalent bonds from atomic structures. Fortunately, most of these energies are known fairly well from experiment, and can serve us well as a starting point. We can at least recognize the interrelationships of various fundamental atomic properties. Then we may make a distinction between individual molecules, or the gaseous state, and nonmolecular solids. It is now possible, as this book will demonstrate, to calculate the individual bond energies in most molecular compounds capable of existence in the gaseous state, and thus to understand their origin in fundamental atomic properties. There still remain relatively few puzzling exceptions, but the available theory is fairly satisfactory for handling most ordinary organic molecules as well as molecular inorganic chemistry. For solids, however, there is usually the additional problem that arises from the structure, which allows the atoms and the bonds they form to be influenced significantly and sometimes in a very complex manner by other atoms which are not directly attached but are close by within the structure. For a fair number of solids, rather simple models permit accurate accounting for their atomization energies. For most solids, however, the exact explanations are elusive, and especially for compounds of the transitional elements.

Throughout the remainder of this book, the emphasis will be on the theory of polar covalence, by which we can understand bond strengths and heats of formation of thousands of major group compounds. However, there will be no intentional omission of areas of ignorance or limitations to the theory. As every good scholar knows, progress in knowledge and understanding is heavily dependent on the recognition of ignorance.

Chapter 2

ATOMIC PROPERTIES AND HOMONUCLEAR BONDING

It is here assumed that the reader is familiar with the electronic configurations of the atoms of the chemical elements. Special attention is directed to the major group elements, wherein the outermost shell never contains more than 8 electrons, because the 9th electron always begins to fill a new higher energy level. The repeated filling of the outermost energy level from 1 to 8 electrons is of course the origin of the periodicity of the elements. All major group elements having the same number of outermost electrons tend to bear a considerable resemblance in other properties, and the properties change with the number of outermost electrons.

Throughout this book all references to periodic groups of elements by number will use the simplified designation of M for major or main group and T for transitional group. The first members are: Li M1, Be M2, Zn M2', B M3, C M4, N M5, O M6, F M7, He M8, Sc T3, Ti T4, V T5, Cr T6, Mn T7, Fe T8, Co T9, Ni T10, and Cu T11.

EFFECTIVE NUCLEAR CHARGE

As mentioned in the preceding chapter, chemical bonding can only occur when the nuclei of both atoms can attract the same electrons. Even though the atoms are electrically neutral, the nuclear charge of one atom must be sensed by electrons of the other atom or else no bond will form. This can come about only when the atom has fewer than 8 outermost electrons, or in other words, when it has orbital vacancies capable of accommodating electrons from another atom. The outer electron of another atom is certainly experiencing a net attraction for its own nucleus or it would not be there. But with respect to the foreign atom, this electron will sense considerable repulsive force between itself and the electrons of the foreign atom, as well as an attraction between itself and the nucleus of the foreign atom. In order to understand the nature and extent of the net force between an outer electron of one atom and the entire other atom, we must consider both the repulsions and the attraction.

It is convenient to picture the repulsions as equivalent to a screening effect which each electron of an atom has upon its own nucleus. That is, each electron is pictured as blocking off a portion of the positive charge of the nucleus. The net nuclear charge, if there is one, that can be sensed within an outer vacancy is the difference between the total nuclear charge and the sum of the screening constants for all the electrons. This difference is called the **effective nuclear charge.** It is always considerably less than the actual nuclear charge, as would be expected considering the repulsive effects of the electrons.

For purposes of this discussion we may be content with the simple rules of John Slater (A1), and in particular, the one which states that electrons in the outermost energy level are very inefficient in blocking off nuclear charge from one another. Clementi and Raimondi (A2) have arrived at screening constants and effective nuclear charges by a much more complex method, but there is no advantage for the present purpose. Each electron in an outermost shell is only about one-third effective in blocking off nuclear charge from the other electrons in that shell. The significance of this is that with **each increase in atomic number,** corresponding to an increase in the number of outermost electrons from 1 to 8, **the effective nuclear charge increases by about two-thirds.** For example, if the effective nuclear charge of sodium is about 2 (approximately true), then that of magnesium is 2 and 2/3, aluminum 3 and 1/3, silicon 4, phosphorus 4 and 2/3, sulfur 5 and 1/3, and chlorine 6. For the outer shell electrons of argon it would be about 6 and 2/3, but this cannot be felt by any electron from another atom because there is no vacancy present into which this electron might be accommodated. This steady increase in effective nuclear charge across the major groups of the periodic table is perhaps the most important consequence of the electronic structure of the atoms, for in turn it profoundly affects other fundamental atomic properties.

ATOMIC RADIUS

One of these properties is the compactness of the electronic sphere, as indicated by the atomic radius. Quantum mechanics denies us the pleasure of treating atoms as solid spheres having hard and definite surfaces, for the probability of finding electrons diminishes only slowly with increasing distance from the nucleus. However, it is useful to define the atomic radius as a **nonpolar covalent radius,** which is half the internuclear distance, or bond length, in a single covalent bond between two like atoms. As the effective nuclear charge increases, the nonpolar covalent radius decreases, changing, for example, from 154 pm for sodium to 99 pm for chlorine, as determined from the experimental bond lengths in Na_2 and Cl_2 molecules. Where the effective nuclear charge remains relatively constant but principal quantum levels are added, the radius increases, as for example, from F 68 to I 133 pm.

ELECTRONEGATIVITY

The electrostatic force of net attraction between an atomic nucleus and an electron from another atom which enters an outer vacancy of the first atom is called the atomic **electronegativity.** It is proportional to the product of the electronic charge and the effective nuclear charge, and inversely to the square of the distance between the charges. From left to right across the major groups of the periodic table, as the number of outermost electrons increases, therefore, the result of **increasing** effective nuclear charge acting over a **decreasing** distance is that electronegativity steadily **increases.** For sodium it is 0.560 and for chlorine it is 3.475, as determined from consideration of the relative compactness (B3, B13) of the electronic spheres of sodium and chlorine. The rationalization is that if the sodium nucleus, which holds a cloud of 11 electrons within a covalent radius of 154 pm, cannot even hold its own electrons closely, it can hardly be expected to attract an electron of another atom very strongly. On the other hand, if the chlorine nucleus holds its cloud of 17 electrons within a radius of only 99 pm, it should be able to attract to its remaining outer vacancy an electron of another atom with considerable force. Electronegativity is, as we shall see in great detail, the **most important property of any atom with respect to its bonding power.**

HOMONUCLEAR BOND ENERGY

Now consider a nonpolar covalent bond, which in its purest form occurs only between like atoms. What is the nature of the force which holds these two atoms together?

Surely it must be closely related to the electronegativity itself, and proportional also to the effective nuclear charge and acting over the distance of the covalent radius, for the average position of the bonding electrons must be midway between the two nuclei. The **homonuclear single covalent bond energy**, required to separate two like atoms held together by this nonpolar covalent bond, must therefore also increase progressively across the major groups of the periodic table. It increases from about 17 kcal (71 kJ) per mole of sodium, Na_2, to 58 kcal (243 kJ) per mole for chlorine, Cl_2.

We have seen that electronegativity, S, is proportional to effective nuclear charge divided by the square of the distance, r^2, and that the homonuclear single covalent bond energy, E, is proportional (A3) to the effective nuclear charge divided by the distance, r, which is the expression for coulombic energy as compared to force. It follows that if two quantities are proportional to the same third, they must be proportional to each other. Therefore E = CrS, where C is a proportionality constant. This relationship reveals which are the most important properties of atoms with respect to their chemical bonds: the **electronegativity,** the **nonpolar covalent radius,** and the **homonuclear covalent bond energy.** Periodic trends in these are represented in Figure 2:1. As will be discussed in detail later, all three of these properties are indispensable in providing quantitative interpretation of bond energies in compounds and consequently of heats of formation and reaction.

EVALUATION OF ELECTRONEGATIVITY

As described in my earlier books, electronegativities can be obtained on the basis of the relative compactness of the electronic spheres of the different atoms, and have also been evaluated in many other ways (B8). Space will not be expended herein to discuss all kinds of electronegativity evaluation because **only the values used herein and no others** have been subjected to the same thorough testing. The values presented here, and those from which these are derived, have been confirmed by their application to calculating the charge distribution and bond energies in many thousands of bonds with very satisfactory accuracy. Many nonpolar single covalent bond energies as well as bond lengths in the elements have been measured very accurately. In order to obtain a complete

Figure 2:1

Typical Trends Across Major Groups of the Periodic Table

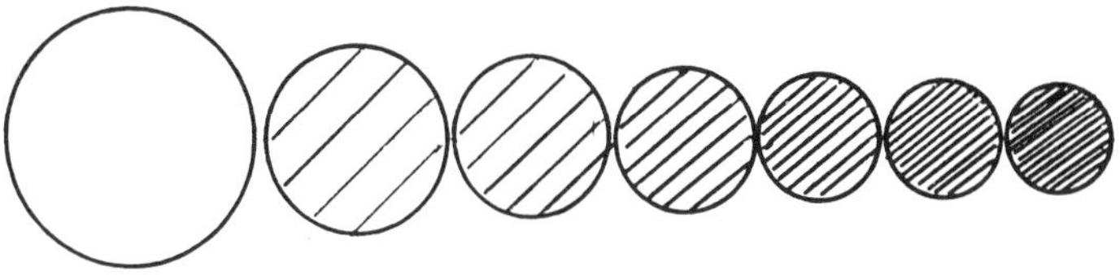

	M1	M2	M3	M4	M5	M6	M7
	(Li)	(Be)	(B)	(C)	(N)	(O)	(F)
no. outer e^-	1	2	3	4	5	6	7
Zeff	2.0	2.7	3.3	4.0	4.7	5.3	6.0
electronegativity	0.670	1.810	2.275	2.746	3.194	3.654	4.000
cov. radius, pm	133.6	88.7	82.2	77.2	73.4	70.2	68.1
BE unweakened	24.6	67.6	76.7	85.4	94.9	104.0	113.1
BE weakened	24.6	67.6	76.7	85.4	38.8	33.6	40.5

set of data which can be used with confidence, it has been necessary to make judicious use of all available experimental data which are relevant and to refine them by noting the nature of certain fundamental interrelationships. A recital of all the details by which the values adopted for this book were obtained would probably prove very boring and certainly not very useful, for they involved numerous intuitive judgments as well as decisions among slightly but significantly conflicting literature values.

However, here are the basic facts. It was found that within each period of major group elements, but excluding the alkali metals and calcium, strontium, and barium, the following properties are linear functions of one another: (1) E and Sr; (2) E and the number of electrons, N, in the outermost shell; (3) E and the function, square root of the atomic number divided by the cube of the radius; (4) the electronegativity S and the reciprocal of the cube of the rradius; (5) Sr and N; and (6) N and $(Z/r^3)^{1/2}$. In particular, it was noted that the straight lines that represent function (4) for the different periods converge when extrapolated to zero radius, at about -0.50 in S. This suggested addition of 0.50 to all electronegativity values. Previous values for the halogens have been the most thoroughly tested (B4), so in this revision it was assumed that their corrected values are accurate. Other electronegativities were adjusted to the same Sr^3 value for that particular period.

For Periods 2 and 3, it was assumed that all revised electronegativities were correct and they were used to revise radius values. For Periods 3 and 4, function (6) above was employed to calculate radii, which were then used to revise electronegativity. The experimental M4 homonuclear energy was used to evaluate the constant, C, in E = CrS, and other values of E were revised using this value of C. Electronegativities of the alkali metals and Ca, Sr, and Ba were obtained from earlier partial charge data and various experimental bond energies.

In the past there seemed to be no need of adopting the Pauling scale for electronegativity evaluation because it was arbitrarily chosen and because of fundamental inaccuracies in the determination of the Pauling values. However, this may well have been a tactical blunder because chemists have become so accustomed to the Pauling scale as to be uncomfortable with a different one. As a concession to conformity, therefore, and in the hope of attracting greater acceptance, since the relative compactness scale was being revised anyway it might just as well be converted to the same range as the Pauling scale, on the basis of a value of 4.000 for fluorine. This required multiplying all the revised values as described above by the factor 0.64. By these procedures was obtained a fully self-consistent set of data capable of successful application to numerous important problems of chemistry. The variations in electronegativity are shown in Figure 2:2. Numerical values are to be found in Table 3:1 of Chapter 3 , where they have more immediate relevance. Figure 2:3 shows how the new electronegativity values compare with the widely accepted values of Allred and Rochow (A3). The relationship between the two sets of values is neither smooth nor linear despite their agreement at zero and 4.000.

Some theorists have insisted that electronegativity is not an atomic property but an orbital property. I have found little or no evidence in all this bond energy work to justify such a distinction. It is, however, important to note that electronegativity, which was originally conceived to

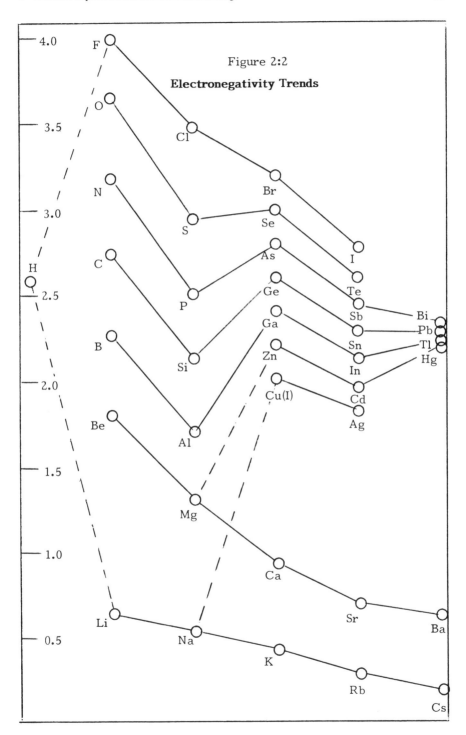

Figure 2:2

Electronegativity Trends

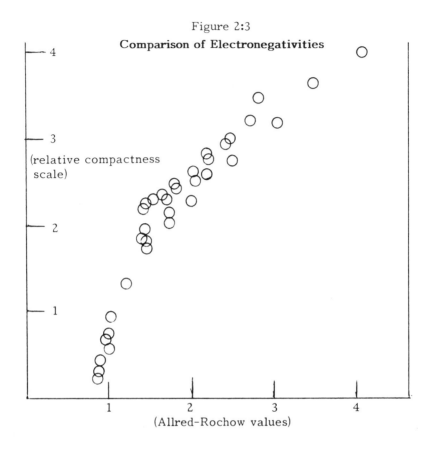

Figure 2:3
Comparison of Electronegativities

(Allred-Rochow values)

account for the greater bond energy of polar bonds, is equally important in determining the energy of nonpolar bonds.

THE LONE PAIR BOND WEAKENING EFFECT

The transition in properties across the major groups of the periodic table from M1 elements, the alkali metals, to M7 elements, the halogens, is the consequence of increasing the number of electrons within the outermost principal quantum level while simultaneously increasing the nuclear charge, as described earlier in this chapter. As a result of the steadily increasing effective nuclear charge, the radius decreases from Li 133.6 to F 68.1 pm. The combination of larger effective nuclear charge acting over a shorter distance corresponds to an increase in electronegativity, from Li 0.670 to F 4.000. Similarly, the homonuclear single covalent bond energy, also a reflection of the effective nuclear charge, increases, from Li 24.6 to carbon 85.4 kcal (102.9 to 357.3 kJ) per mole. Beyond carbon, the effective nuclear charge continues to increase, the radius continues to decrease (from C 77.2 to F 68.1 pm), and the electronegativity continues to increase (from 2.746 to F 4.000). A corresponding increase in homonuclear bond energy is logically anticipated. In fact, linear extrapolation of the experimental homonuclear single covalent bond energy from Li through C as a function of the product of radius and electronegativity gives the following energies: N–N 94.9 (397), O–O 104.0 (435), and F–F 113.1 (473) kcal (kJ) per mole. The trend is as predicted.

Experimental determinations, however, provide energies much lower than these. For example, there seems to be no obvious reason for any very significant difference in bond energy between N–H in ammonia and N–H in hydrazine, NH_2–NH_2. The average bond energy in ammonia is 93.4 kcal (390.8 kJ). If four times this energy is subtracted from the experimental atomization energy of hydrazine, which is 412 kcal (1724 kJ) per mole, only about 38 kcal (159 kJ) is left for the N–N bond. Similarly, if the O–H bond energies in water are assumed to be essentially the same as in hydrogen peroxide, HO–OH, and their value of 221.6 kcal (927.1 kJ) is subtracted from the total atomization energy of the hydrogen peroxide, which is 255.9 kcal (1070.7 kJ) per mole, this leaves only about 34 kcal (142 kJ) for the O–O bond. Finally, the low dissociation energy of F_2, about 38 kcal (159 kJ) per mole, was controversial for a long period because it was expected to be substantially higher than the 58 kcal (242.7 kJ) per mole observed for Cl_2.

Nitrogen

To understand these large anomalies, examination of the multiple bonding of nitrogen and oxygen should be useful. The bond energy in nitrogen, N_2, is 225.9 kcal (945.2 kJ) per mole, compared with about 196 kcal (820 kJ) for the carbon–carbon triple bond in alkynes. This is consistent with the upward trend expected for single bond energies, but it is nearly six times the N–N energy obtained from hydrazine, whereas in alkynes a triple bond is only about 2.4 times as strong as a single bond. As detailed in Chapter 4, multiple bond energies are found by multiplying the single bond energy at the appropriate (experimental) bond length by an empirical factor, which is 1.787 for triple bonds and 1.488 for double bonds. From the triple bond factor, the experimental bond length of 109.8 pm, and the covalent radius sum of 146.8 pm which would be the expected bond length of a nonpolar single N-N bond, we may calculate what the single bond energy should be to correspond to the observed triple bond energy in N_2:

$$E = (225.9 \times 109.8)/(1.787 \times 146.8) = 94.6 \text{ kcal } (395.8 \text{ kJ})$$

Since this is nearly identical with the extrapolated single bond energy of 94.9 kcal (397.1 kJ) per mole for N-N, it seems reasonable to conclude, at least tentatively, that the homonuclear single covalent bond energy for nitrogen should indeed be close to 94.9. If so, the very low value of 38 in hydrazine must reveal some **unsuspected weakening effect.**

The only unusual happening at nitrogen, or M5, in crossing the major groups from lithium to fluorine, is the **first appearance of a lone pair of electrons** in the outermost principal quantum level. If these lone pair electrons are responsible for the weakening effect, formation of a triple bond, which concentrates six electrons within the internuclear region, must repel the lone pair electrons to the far side of the nitrogen atoms where their weakening action evidently cannot function. Until proven innocent, the lone pair electrons appear guilty of the weakening effect, so it is named the **lone pair bond weakening effect (LPBWE).** It is observed only in the chemistry of the major group elements from M5, the nitrogen group, to M7, the halogens, these being the only major group elements whose atoms have lone pair electrons in their valence shell.

Herein lies a basis for explaining why nitrogen exists as N_2 molecules instead of the more reasonably expected solid polymeric form analogous to phosphorus. The LPBWE is so large for nitrogen that triple bonds, which eliminate it, are clearly favored over their equivalent in single bonds, which are fully weakened. Atomization of the solid would require 1.5 x 38.8 = 58.2 kcal (243.5 kJ) per mole of atoms, since breaking of each of three bonds would liberate two atoms from that bond. Atomization of N_2, on the other hand, requires 113 kcal (473 kJ) per mole of atoms. N_2 is 55 kcal (230 kJ) per mole of atoms more stable than the solid polymer would be. In a similar manner it has been shown (B4) that forms of nitrogen in which it has alternate double and single bonds, in open or closed chains, or analogous to benzene, a ring of nitrogen atoms, N_6, although more stable than the single-bonded polymer, are still appreciably less stable than N_2.

Oxygen

For oxygen, the bond energy in O_2 is 119.1 kcal (493.2 kJ) per mole, about 3.5 times larger than the single bond energy obtained from hydrogen peroxide. This is very different from the situation in hydrocarbons, wherein the double bond energy, about 145 kcal (607 kJ) per mole, is only about 1.75 times as large as the single bond energy of about 83 kcal (347 kJ) per mole. Yet the O_2 value is considerably smaller than that of the C=C bond although it would be expected to be larger. Again we may calculate what a single bond energy should be, this time to correspond with the double bond energy of 119.1, by using the double bond factor of 1.488 and the single nonpolar bond length of 140.4 and the experimental value of 120.7 pm.

$$E = (119.1 \times 120.7)/(1.488 \times 140.4) = 68.8 \text{ kcal } (287.9 \text{ kJ})$$

From the fact that the O_2 bond energy is less than a C=C bond energy although much larger than the weakened single bond energy, it may be assumed that in forming a double bond, the oxygen single bond energy remains partly weakened but the weakening is reduced. If it were reduced by half, then the difference of 35 kcal between the 69 and the 34 could be added to the 69 to give 104 kcal (435 kJ) for the completely unweakened bond energy. This is identical with the extrapolated value given earlier. It may tentatively be concluded that the 104 kcal represents the unweakened homonuclear single covalent bond energy of oxygen, 69 the half-weakened, and 34 the fully weakened energy. If the unweakened energy were involved in the double bond, assuming the experimental bond length to be the same, the O=O energy would be calculated as 180 kcal (753 kJ) per mole, about what might be expected for a logical increase from the C=C energy of about 145 kcal (607 kJ).

The existence of oxygen as gaseous O_2 molecules instead of a polymeric solid held together by single bonds can be explained in a manner similar to that used for N_2. Although the LPBWE is reduced only by half in the formation of a double bond, this reduction is large enough in oxygen to favor double bonds over their equivalent in single bonds. Atomization of the solid polymer would require 33.6 kcal (140.6 kJ) per mole of atoms, since each bond broken liberates two atoms from it. Atomization of O_2 requires 59.6 kcal (249.4 kJ) per mole of atoms. The gas is 26 kcal (109 kJ) per mole of atoms more stable than the solid would be.

The great importance of the LPBWE is well illustrated by these two examples of nitrogen and oxygen. Without this effect, life as we know it would be impossible. Intake of air would require chewing instead of breathing!

Other Elements

Extrapolation in the other periods similarly led to evaluation of unweakened energies for the heavier elements of these groups. For reasons to be discussed presently, atoms of these elements tend to join together by single rather than multiple bonds. These are the fully weakened bonds, providing us with experimental values for most of the fully weakened bond energies. All M5-M7 energies are given in Table 3:1. It is convenient to differentiate the three homonuclear energies of each element by designating the unweakened value as triple prime, E''', the half-weakened value as double prime, E'', and the fully weakened value as single prime, E'.

The unweakened homonuclear energies are shown in Figure 2:4. The same, but fully weakened in M5, M6, and M7, are shown in Figure 2:5. A careful comparison of the two figures should provide confidence that the unweakened energies are at least reasonable and orderly. It would not be easy to make sense of the experimental data of Figure 2:5 without a quantitative understanding of the LPBWE.

The exceptional reactivity of fluorine which makes it a sort of "superhalogen" is clearly not merely the result of the fact that fluorine is the most electronegative of all the chemical elements. It reflects also the great ease of atomization of F_2 molecules to provide the highly reactive free atoms. The long puzzling anomaly of the unexpectedly

Figure 2:4

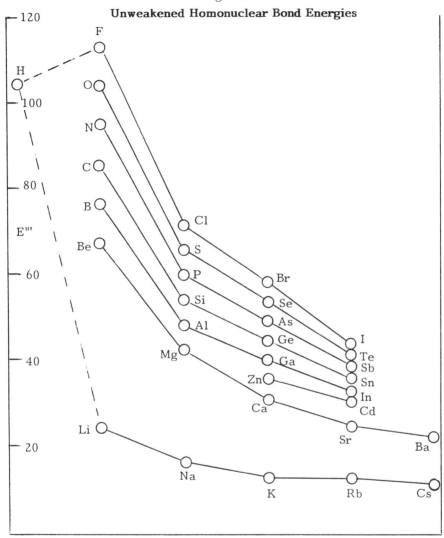

Unweakened Homonuclear Bond Energies

NOTE: These are the energies expected if there were no LPBWE. Compare them with the values in Figure 2:5, which are the same for Groups M1-M4, but erratically different from those of M5-M7. Unweakened, the bonds exhibit a reasonably regular pattern, as expected from regularities in atomic structure.

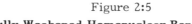

Figure 2:5

Fully Weakened Homonuclear Bond Energies

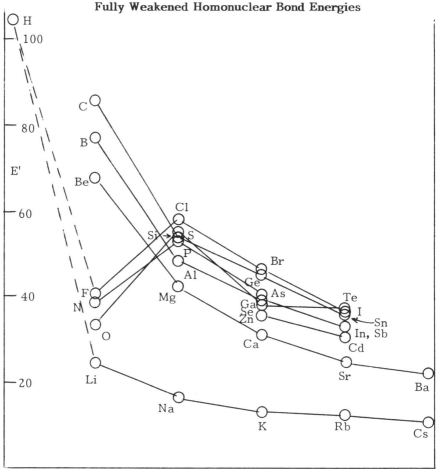

NOTE: These bond energies are the same as in Figure 2:4 for Groups M1-M4, but exhibit the LPBWE in Groups M5-M7. The trends are therefore wholly inconsistent with expectations from regularities in atomic structure. Pending quantitative explanation of LPBWE, no simple pattern can be discerned. See also Figure 5:1.

low dissociation energy of F_2 compared with the other halogens is now easily explicable, to the extent that the large LPBWE is recognized. Note (Table 3:1) the trend in E''' values for F_2 113.1, Cl_2 71.8, Br_2 58.3, and I_2 45.9 kcal per mole is just what one might expect. It is only the **weakened,** which are the actual bond dissociation energies, that are **anomalous.**

The existence of the heavier elements in the solid instead of gaseous state reflects the relatively small lone pair bond weakening effect in these elements, such that multiple bonds are not favored over their equivalent in single bonds. For example, it is sometimes thought that phosphorus atoms exist as solid, joined together by single bonds, because they cannot form triple bonds. In fact, when phosphorus vapor, normally P_4, is heated above 800°C it becomes P_2 molecules, which closely resemble N_2. The bond length is 189 compared with 221.4 pm for the single bond, and the dissociation energy is 125.1 kcal (523.4 kJ) per mole. This dissociation energy can be calculated from the extrapolated unweakened E''' value for P which is 60.0 kcal (251 kJ) per mole:

$$60.0 \times 1.787 \times 221.4/189 = 125.6 \text{ kcal } (525.5 \text{ kJ})$$

Or, the E''' energy could be calculated from the experimental dissociation energy:

$$125.1 \times 189 = 1.787 \times 221.4 \text{ E'''}; \quad \text{E'''} = 59.8 \text{ kcal } (250 \text{ kJ})$$

To atomize P_4, three bonds per phosphorus atom must be broken but each break liberates 2 atoms from that bond. Therefore the atomization energy is 1.5 times the single bond energy of 53.7, or 80.6 kcal (337.2 kJ) per mole of P atoms. However, to atomize P_2 requires only half the energy of dissociation per mole of atoms, or 62.8 kcal (261.9 kJ). This is why the solid P_4 state is preferred over the gaseous P_2 at ordinary temperatures. Similar calculations can be performed to show that arsenic and antimony, both capable of forming diatomic molecules triply bonded, at high temperatures, are more stable as singly-bonded solids at ordinary temperatures.

In a similar manner the solid condition of sulfur can be examined. Between 500 and 1900°C the principal species

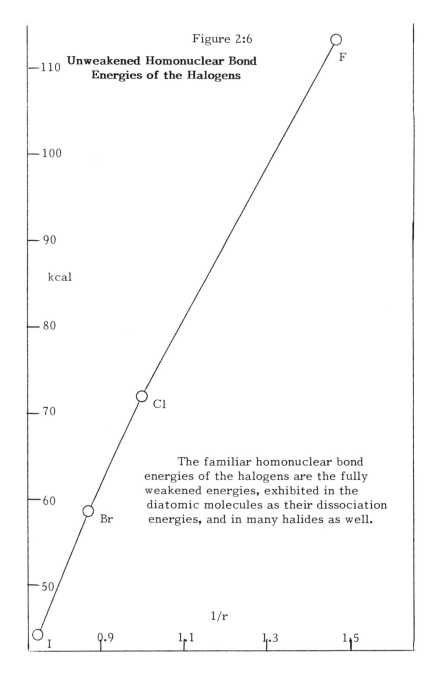

Figure 2:6

**Unweakened Homonuclear Bond
Energies of the Halogens**

The familiar homonuclear bond
energies of the halogens are the fully
weakened energies, exhibited in the
diatomic molecules as their dissociation
energies, and in many halides as well.

in sulfur vapor is S_2, which closely resembles O_2 in being paramagnetic and having a double bond. The dissociation energy experimentally determined is 102.5 kcal (428.9 kJ) per mole. The bond length is 189 pm. The E" energy for sulfur is 60.4 kcal (252.7 kJ), and the nonpolar covalent radius sum, or single bond length is 209.8 pm. The energy of a double bond can be calculated:

$$1.488 \times 60.4 \times 209.8/189 = 99.8 \text{ kcal } (417.4 \text{ kJ})$$

The S' single bond energy is 54.9 kcal (229.7 kJ), and this would be the atomization energy of singly-bonded sulfur, compared with only half of 102.5 or 51.3 kcal (214.6 kJ) for the atomization of S_2. The advantage of the solid state is clear, although small. However, another factor enters into the bonding in solid sulfur, which is most stable in the form of S_8 rings. The experimental atomization energy of sulfur is 66.6 kcal (278.7 kJ) per mole. From the heat of vaporization of S_8, 24.5 kcal (102.5 kJ) per mole, one-eighth or 3.1 kcal (13.0 kJ) per sulfur atom is required to overcome the van der Waals attractions among molecules. Therefore the average bond energy in S_8 vapor is 66.6 - 3.1 or 63.5 kcal (266.7 kJ). This is intermediate between the E" and E''' values in Table 3:1, 60.4 kcal (252.7 kJ) and 65.9 kcal (275.7 kJ) per mole, which suggests that for some reason not immediately obvious, the lone pair weakening effect is largely absent in S_8. Whatever the reason, there is greater weakening in other molecular species so S_8 is favored. This gives the solid state a substantially larger advantage over S_2 gas than was calculated for the fully weakened S–S bonds.

The lone pair bond weakening effect is very much involved in compounds of M5, M6, and M7 elements as well as in these elements themselves, as will be discussed in detail later.

ATOMIZATION ENERGIES OF THE ELEMENTS

Before determining experimental atomization energies of chemical compounds from their standard heats of formation, we must know accurately the atomization energies of the individual elements. This is because the elements in their standard states consist of atoms already bonded together, and these bonds must be broken before the atoms can

rearrange to form the compound. Therefore the heat of formation of the compound reflects not only the energy released
when the atoms form the compound, but also the energy
expended when the atoms were liberated from the standard
state of the element. **The atomization energy of a compound
is therefore the difference between the sum of the atomization
energies of the component elements and the standard heat
of formation of the compound.** The atomization energies
used in this work were the most accurate available in standard
sources in the literature (A4-A6). They are given in Table
2:1.

MORE MEANINGFUL THERMOCHEMISTRY

Standard practice in the presentation of thermochemistry to students has for many years involved standard heats
of formation, with the convention that these heats are zero
for the elements in their standard (common) states at 25°C
and one atmosphere pressure. This practice has been consistent with the tabulation of standard heats of formation by
various groups such as the National Bureau of Standards (A5),
CODATA (A37), the Joint Army, Navy, and Air Force (JANAF)
compilers (A12), and in collections such as that of Cox and
Pilcher (A13). This practice has the advantage that knowledge
of atomization energies of the individual elements has not
been required.

However, it is time for a change. The purpose of this
section is to suggest such a change and explain its advantages
in the teaching of chemistry.

The atomization energies of the chemical elements
(Table 2:1) are now reasonably well known. The change I
suggest is to use atomization energies of elements and compounds in place of standard heats of formation wherever
possible. The atomization energy of a compound provides
directly the sum of the contributing bond energies. For example, the standard heat of formation of $H_2O(g)$ is -57.8 kcal
(-241.8 kJ) per mole, and the heats of atomization of two
hydrogen atoms, 104.2 kcal (436.) kJ), and one oxygen atom,
59.6 kcal (249.4 kJ), add up to 163.8 kcal (685.3 kJ). The
atomization energy of $H_2O(g)$ is 163.8 minus -57.8 or 221.6
kcal (927.1 kJ) per mole, this being the total strength of the
two O-H bonds.

Table 2:1

Atomization Energies of the Chemical Elements

element	at. energy		element	at. energy		element	at. energy	
	kcal	kJ		kcal	kJ		kcal	kJ
Ag	68.1	284.9	H	52.1	218.0	Re	185.	774.
Al	78.8	329.7	Hf	148.	619.	Rh	153.	557.
As	72.3	302.5	Hg	14.7	61.5	Ru	155.5	648.5
Au	88.0	368.2	I	25.5	106.8	S	66.2	277.0
B	133.8	559.9	In	58.	243.	Sb	63.2	264.4
Ba	42.5	177.8	Ir	160.	669.	Sc	90.3	377.8
Be	77.5	324.3	K	21.4	89.6	Se	54.3	227.2
Bi	50.1	209.6	Li	38.6	161.5	Si	107.6	450.2
Br	26.7	111.9	Mg	35.2	147.3	Sm	49.4	206.7
C	171.3	716.7	Mn	67.7	283.3	Sn	72.0	302.1
Ca	42.6	178.2	Mo	157.3	658.1	Sr	39.1	163.6
Cd	26.7	111.8	N	113.0	472.7	Ta	186.9	782.0
Ce	101.	423.	Na	25.8	108.1	Te	47.0	196.7
Cl	29.0	121.3	Nb	172.4	721.3	Th	142.9	597.9
Co	102.4	428.4	Ni	102.8	430.1	Ti	112.3	469.9
Cr	95.	398.	O	59.6	249.2	Tl	43.6	182.2
Cs	18.7	78.2	Os	188.	787.	U	126.	527.
Cu	80.7	337.6	P	75.6	316.3	V	122.9	514.2
Er	75.8	317.1	Pb	46.7	195.1	W	203.1	849.8
F	18.9	79.1	Pd	90.0	376.6	Y	101.5	424.7
Fe	99.3	415.5	Pt	135.2	565.7	Yb	36.4	152.1
Ga	65.4	273.6	Pu	87.1	364.4	Zn	31.2	130.4
Ge	89.5	374.5	Rb	19.6	82.0	Zr	145.5	608.8

Given only standard heats of formation, students are likely to get the idea that positive values mean weak, unstable bonds. This is of course not necessarily true, for a positive value of $\Delta Hf°$ means only that the bonds are not as strong as in the free elements, by that amount. For example, for NO(g), $\Delta Hf°$ is +21.6 kcal (90.4 kJ) per mole, yet the bond energy is 151.0 kcal (631.8 kJ) per mole. This bond is even stronger than in HF where $\Delta Hf°$ is -64.8 kcal (-271.1 kJ) per mole, corresponding to a bond energy of 135.8 kcal (568.2 kJ). For Cl_2O(g), $\Delta Hf°$ is almost the same as for NO(g), +19.2 kcal (80.3 kJ), corresponding to an atomization energy of 98.4 kcal (411.7 kJ) per mole, only about two-thirds that of NO.

One might also think comparisons of standard heats of formation would be meaningful. They are not. For example, $\Delta Hf°$ for $CaCl_2$(c) is -190.2 kcal (-795.8 kJ) per mole, but $\Delta Hf°$ for WCl_2(c) is only -61 kcal (-255 kJ) per mole. $CaCl_2$ is obviously much more stable? Not so. The atomization energy of WCl_2(c) is 322.1 kcal (1347.7 kJ) whereas it is only 290.8 kcal (1216.7 kJ) for $CaCl_2$(c).

When students are taught to use standard heats of formation to determine heats of reaction, and vice versa, they use mysterious numbers from a table, find the signs confusing, and with luck obtain a correct answer with no idea of why, and very little notion of what the answer means. On the other hand, if they began with tabulated atomization energies, all values would be positive and they would arrive at the same answer but understand its meaning. Atomization energy is the total bond strength within the molecule. All chemical reactions are rearrangements of atoms, breaking former bonds to make new ones, and they tend to proceed in the direction of forming stronger bonds. The heat of reaction is simply the difference between the total atomization energy of the reactants and that of the products.

For example, consider the reaction:

	CH_4	+	Cl_2	=	CH_3Cl	+	HCl	ΔH of reaction
$\Delta Hf°$	-17.9		0		-19.3		-22.1	-23.5
ΔH_{at}	397.6		58.0		375.9		103.2	-23.5

4 C–H   Cl–Cl   3 C–H   H–Cl

C–Cl

From heats of formation, the student learns nothing but the heat of reaction. From atomization energies he sees what bonds are broken at what cost and what new bonds form at what gain, and understands the **origin** of the heat of reaction.

Or, consider the heat of combustion of carbon, the most important reaction for the production of useful energy. Since this is the same as the heat of formation of CO_2, students can find it in a table, -94.1 kcal (-393.7 kJ) per mole. But what meaning does this have? If they use heats of atomization,

	C	+	O_2	=	CO_2	ΔH of reaction
ΔH_{at}	171.3		119.1		384.5	-94.1 kcal
	graphite		O=O		2 C=O	

they can see that energy is required to atomize the carbon, breaking the bonds in graphite, and to atomize the O_2 into two O, totalling 290.4 kcal (1215.0 kJ). This much must be expended per mole of CO_2 to be formed. If more than that is not evolved when the carbon atoms combine with the oxygen atoms, no heat will be released. Fortunately the bond strength in CO_2 is 94.1 kcal (393.7 kJ) greater than in the corresponding graphite and O_2, so this much heat is evolved.

Finally, consider the synthesis of hydrogen chloride by reaction of sulfuric acid with sodium chloride:

	H_2SO_4(liq)	+	NaCl(c)	=	HCl(g)	+	$NaHSO_4$	ΔH of reaction
$\Delta Hf°$	-194.5		-98.3		-22.1		-269.0	1.7
ΔH_{at}	603.1		153.2		103.2		651.4	1.7
								kcal

Although most commonly, reactions tend to occur in the direction of greater bond strength, here the heat of reaction, which is positive, is so small that the entropy change is sufficiently positive to make the free energy change negative, according to the expression: $\Delta G = \Delta H - T \Delta S$. This entropy change can easily be predicted by observation that the reaction liberates a gas, thereby increasing the entropy. Therefore the reaction will proceed "spontaneously."

Consider also the second step:

	$NaHSO_4(c)$	$+$ $NaCl(c)$	$=$ $HCl(g)$	$+$ $Na_2SO_4(c)$	ΔH of reaction
$\Delta Hf°$	-269.0	-98.3	-22.1	-331.5	13.7
ΔH_{at}	651.4	153.2	103.2	687.7	13.7

Here the increase in entropy is insufficient to make up for the positive heat of reaction and the mixture must be heated to bring about the reaction. These explanations can be confirmed by substitution of free energy for heat or enthalpy.

	$H_2SO_4(liq)$	$+$ $NaCl(c)$	$=$ $HCl(g)$	$+$ $Na_2SO_4(c)$	ΔG of reaction
$\Delta Gf°$	-164.9	-91.8	-22.8	-272.3	-3.4
ΔG_{at}	540.6	135.5	96.7	582.8	-3.4

	$NaHSO_4(c)$	$+$ $NaCl(c)$	$=$ $HCl(g)$	$+$ $Na_2SO_4(c)$	
$\Delta Gf°$	-237.3	-91.8	-22.8	-303.6	2.7
ΔG_{at}	582.8	135.5	96.7	618.9	2.7

Again, the main point is that the atomization energies--free energies or enthalpies--are much more meaningful than the standard values of $\Delta Hf°$ or $\Delta Gf°$.

Similarly, the common practice of trying to present electrode potentials without describing them as thermochemical quantities has always seemed unwise. Even though they involve free energies rather than enthalpies, they can be quantitatively explained from a thermochemical cycle involving free energies of atomization, ionization or electron affinity, and hydration energy of the ions (B36). For example, it can easily be shown (but seldom is) that the principal origin of the difference in electrode potential between copper and zinc is the much greater bonding energy in metallic copper.

In summary, why continue a less helpful practice just because it is of long standing? Custom should be a useful guide, not a prison. The theory of polar covalence, as will be seen, can contribute much to thermochemistry by explaining atomization energy. The latter can be much more useful than heat of formation in making thermochemistry more meaningful.

Chapter 3

THE THEORY OF POLAR COVALENCE

As Pauling first supposed (A7), when two different kinds of atom join by a covalent bond, the energy of this bond should be an average of the two homonuclear bond energies. In other words, as long as the electrons are evenly shared, atom A will contribute a fixed amount to the total energy regardless of the identity of atom B. However, different kinds of atoms are almost certain to differ in their attraction for electrons. This results in partial charges on the combined atoms which must influence the total energy of the bond, by producing an ionic character to strengthen the bond. Pauling assumed, as the basis for his evaluation of electronegativity, that this ionic contribution is the difference between the experimentally observed bond energy and the energy if the bond were nonpolar covalent. However, in a nonpolar covalent bond the two electrons are shared evenly and in an ionic bond they are monopolized by the initially more electronegative atom. **If they are shared evenly all of the time they cannot be unevenly shared part of that time.** In other words, any degree of ionicity in the bond must **be deducted from the total covalence.** The total bond energy therefore cannot accurately be represented as the sum of the total possible covalent energy plus an ionic energy, but rather, as **the sum of the ionic energy and that part of the covalent energy which remains after the degree of covalence has been reduced by substitution of an ionic contribution.** This concept **decreases** the contribution to total bond energy made by **covalence** and **increases** the contribution made by **ionicity,** compared with Pauling's original suppositions. The identical total bond energy is divided differently into covalent and ionic portions.

NONPOLAR COVALENT ENERGY

The hypothetical covalent energy if a polar covalent bond were nonpolar can be represented as the geometric mean of the two homonuclear bond energies, as suggested by Pauling, but corrected for any difference between the actual bond length and the sum of the two nonpolar covalent radii. This is because of the inverse dependence of bond

35

energy on internuclear distance and because the homonuclear bond energies are valid only for the length of the nonpolar covalent radius. Bond polarity usually shortens the bond to less than the sum of the nonpolar covalent radii, although changes in both directions are observed and must be corrected for. We therefore can calculate the total covalent energy, E_C, as follows:

$$E_c = R_o(E_{AA}E_{BB})^{1/2}/R_o$$

E_{AA} and E_{BB} are the homonuclear single covalent bond energies of atoms A and B, R_C is the sum of the two nonpolar covalent radii, and R_o is the experimental bond length.

IONIC ENERGY

Representation of the total energy if it were all ionic is even simpler. It is, for a diatomic molecule, the product of unit charges divided by the distance between them, or bond length:

$$E_i = 33200/R_o$$

where 33200 is the factor needed to convert the units to kcal per mole if R_o is in pm, or the factor is multiplied by 4.184 if the energy is to be expressed in kiloJoules.

Note that the factor 33200 is divided by R_o which seldom is larger than 300 pm and never larger, in a molecule, than 332 (CsI g). Therefore the ratio is never less than 100. In contrast, there is no geometric mean homonuclear energy for any bond that is as large. Consequently the ionic contribution to polar covalent bond energy must always be larger than that part of the covalent energy which it replaces. This is why polarity **always** increases bond strength.

POLAR COVALENT ENERGY

The problem in developing a satisfactory theory of polar covalence is to describe the actual bond which is intermediate between these extremes. We may begin by considering the nature of uneven sharing of electrons. The simplest example is that of a diatomic molecule, AB, formed by a single covalent bond between atoms A and B. Let us assume that atom B is more electronegative than atom A. The two electrons will initially be attracted more strongly by the nucleus of B than by the nucleus of A. This will cause them to spend more than half time around nucleus B, imparting a partial negative charge to atom B and leaving the atom A with a partial positive charge. Remember that the origin of the attraction between nucleus and bonding electrons is the existence of nuclear charge within the bonding orbitals, that is effective despite the repulsions of the other electrons. Remember too that the effective nuclear charge is determined by both the actual nuclear charge and the shielding of the nucleus by the electrons. When atom B acquires a partial negative charge, this means, in effect, an increase in its electronic population, which will very significantly influence the nature of the atom. By increasing the repulsions it will cause the electronic sphere to expand. By increasing the shielding of the nucleus it will decrease the effective nuclear charge. This **reduction** of effective nuclear charge combined with **increase** in the distance over which it acts corresponds to a **decrease in the electronegativity** of atom B.

At the same time, the removal of part of an electron from the sphere of atom A will also have very significant effects. By diminishing the electronic population around the nucleus of atom A it will reduce the interelectronic repulsions and allow the electronic sphere to contract. It will also reduce the shielding of the nucleus, allowing an increase in the effective nuclear charge. An **increased** effective nuclear charge acting over a **shorter** distance corresponds to an **increased electronegativity.** So, while the electronegativity of B **decreases,** the electronegativity of A **increases.** These adjustments must cease when the two atoms have become **equal in electronegativity.**

Principle of Electronegativity Equalization

Similar changes, except as noted later, occur whenever atoms join together to form molecules or nonmolecular solids. This tendency is expressed as the **Principle of Electronegativity Equalization.** It was first published in 1951 (B12), proven correct by quantum mechanics in 1978 by Robert Parr and coworkers (A8), and confirmed in 1979 by Peter Politzer and H. Weinstein (A9). It may be stated: **When two or more atoms unite to form a compound, their electronegativities become adjusted to the same intermediate value within the**

Figure 3:1

Representation of Electronegativity Equalization

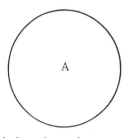

Atom A has relatively few outer electrons, corresponding to:

1) a relatively **small** effective nuclear charge
2) a relatively **large** nonpolar covalent radius
3) a relatively **diffuse** electronic cloud

and therefore 4) a **low electronegativity.**

Atom B has a nearly filled outer octet, corresponding to:

1) a relatively **large** effective nuclear charge
2) a relatively **small** nonpolar covalent radius
3) a relatively **compact** electronic cloud

and therefore 4) a relatively **high** electronegativity.

When the two atoms join by a covalent bond, A-B, the bonding electrons are attracted initially more strongly to B, giving it a surplus (or partial negative charge), and leaving A with a deficiency of electrons (or partial positive charge).

With its electron population **reduced,**
　　a) the effective nuclear charge of A **increases**
　　b) causing the cloud to **contract** to **smaller radius**
　　c) and A to be **more electronegative.**

With an increase in electron population,
　　a) the effective nuclear charge of B **decreases**
　　b) causing the cloud to **expand** to **larger radius**
　　c) and B to be **less electronegative.**

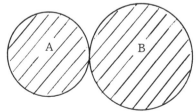

These adjustments cease when the two atoms have become **equal** in **electronegativity,** by virtue of acquiring **partial charge.**

compound. A corollary is, **the intermediate electronegativity within the compound is the geometric mean of all the atomic electronegativities.**

Note that the different kinds of atoms become **equal** in electronegativity by **unequal** sharing of the bonding electrons. In the example of the molecule A–B (Figure 3:1), the bonding electrons can be pictured as spending more than half time around the nucleus of atom B, but as being attracted just as strongly to the nucleus of A during the less than half time they are more closely associated with that nucleus.

Model of Polar Covalence

This undoubtedly affects the energy of the bond in a complex manner. However, although it is not clear how best to describe the real bond, we can approach a reasonable description by recognizing that it is intermediate between two precisely definable and describable hypothetical extremes, **nonpolar covalence** and **ionicity.** There are so many distinguishable shades of grey between white and black that it would be difficult to provide a unique description to a particular shade. However, we know that all these shades may be produced by blending appropriate relative amounts of white and black. The theory of polar covalence suggests that the real polar covalent bond can be adequately described, from the viewpoint of its actual energy, as a **weighted blend** of a **covalent** contribution and an **ionic** contribution, just as a shade of **grey** paint can be quantitatively specified as the result of mixing fixed quantities of **white** with **black.**

According to this theory, the actual energy of a polar covalent bond can be expressed:

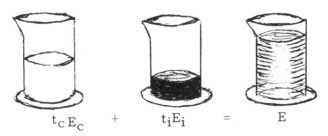

$$t_c E_c \quad + \quad t_i E_i \quad = \quad E$$

This energy is the "contributing bond energy (CBE)", defined as that part of the total atomization energy which that bond contributes. Except in diatomic molecules, this is not the same as the "bond dissociation energy (BDE)" described in Chapter 12. The blending coefficients are t_c, the covalent blending coefficient, and t_i, the ionic blending coefficient. Their sum is 1.000. The method of calculating E_c and E_i was previously shown. Now the problem is the evaluation of the blending coefficients. Fortunately this is quite easy. The ionic blending coefficient is equal to half the difference between the two partial charges, and the covalent blending coefficient, t_c, = 1.000 – t_i.

Partial Charge

This brings us to the evaluation of partial charge, which can also be done quite simply, as described about 30 years ago. This was originally based on two assumptions: (1) the electronegativity of a combined atom changes linearly with the partial charge on the atom; and (2) the bond in an isolated molecule of NaF is 75 per cent ionic (since revised to about 80 per cent). The change in electronegativity that corresponds to acquisition of unit charge, positive or negative, ΔS_i, was evaluated and found to be equivalent to a constant, 2.08, times the square root of the atomic electronegativity:

$$\Delta S_i = 2.08 \ S^{1/2}$$

This constant, 2.08, is changed, however, by the recent revisions, to 1.56: ΔS_i = 1.56 $S^{1/2}$. The **partial charge** is then defined as the **ratio of the change in electronegativity undergone by an atom in forming the compound, to the change that would correspond to acquisition of unit charge.** The data needed for calculation of bond energies in molecular compounds are summarized in Table 3:1.

An Example: Gaseous KBr

An example of a polar covalent bond should be helpful here. Immediately we are faced with a problem. Practically all chemical compounds which can be studied in the gaseous state are either compounds of metal with nonmetal or non-metal with nonmetal. All the nonmetals beyond M4 are subject to lone pair bond weakening. The problem is, which homo-nuclear bond energy is appropriate for the nonmetal? This problem will be thoroughly discussed in Chapter 5. For our example, let us take a gaseous molecule of potassium bromide, KBr, and let us accept the dissociation energy of Br_2 as the appropriate homonuclear energy. This happens to be the

Table 3:1

Data for Calculating Polar Bond Energies

	r, pm	S	ΔS_i	At. E	E'''	E''	E'
H	32.	2.592	2.528	52.1	104.2		
Li	133.6	0.670	1.285	38.6	24.6		
Be	88.7	1.810	2.112	77.5	67.6		
B	82.2	2.275	2.368	133.8	76.7		
C	77.2	2.746	2.602	171.3	85.4	(82.8 n-alk.	
N	73.4	3.194	2.806	113.0	94.9	66.9	38.8
O	70.2	3.654	3.001	59.6	104.0	68.8	33.6
F	68.1	4.000	3.140	18.9	113.1	76.8	40.5
Na	153.9	0.560	1.175	25.8	16.4		
Mg	137.3	1.318	1.804	35.2*	43.3		
Al	125.8	1.714	2.055	78.8	48.2		
Si	116.9	2.138	2.296	107.6	54.1		
P	110.7	2.515	2.490	75.6	60.0	56.9	53.7
S	104.9	2.957	2.700	66.2	65.9	60.4	54.9
Cl	99.4	3.475	2.927	29.0	71.8	64.9	58.0
K	196.2	0.445	1.047	21.4	13.1		
Ca	174.	0.946	1.527	42.6	30.8		
Cu	133.1	2.033	2.239	80.7	31.3		
Zn	129.2	2.223	2.341	31.2	35.8		
Ga	125.6	2.419	2.442	65.4	40.3		
Ge	122.3	2.618	2.540	89.5	44.8		
As	119.4	2.816	2.635	72.3	49.3	45.1	40.9
Se	116.7	3.014	2.726	54.3	53.8	46.0	38.2
Br	114.2	3.219	2.817	26.7	58.3	52.2	46.1
Rb	216.	0.312	0.887	19.6	12.4		
Sr	191.	0.721	1.333	39.1	24.6		
Ag	153.3	1.826	2.122	68.1	27.7		
Cd	149.3	1.978	2.208	26.7	30.4		
In	145.5	2.138	2.296	58.	33.1		
Sn	142.0	2.298	2.380	72.0	35.8		
Sb	138.9	2.458	2.461	63.2	38.5	35.8	33.0
Te	136.0	2.618	2.540	47.0	41.2	39.2	37.2
I	133.3	2.778	2.617	25.5	43.9	40.0	36.1
Cs	235.	0.220	0.376	18.7	10.8		
Ba	198.	0.651	1.267	42.5	22.2		
Hg	150.0	2.195	2.326	14.7	8.6		
Tl	149.0	2.246	2.353	43.6	16.6		
Pb	148.0	2.291	2.376	46.7	24.2		
Bi	147.0	2.342	2.403	50.1	32.2		

*experimental value; empirical 50.3 preferred.

E''' unweakened, E'' half weakened, E' fully weakened bond energy.

fully weakened energy, but certainly it is the familiar one. For potassium, we have also an experimental value for the dissociation energy of K_2 molecules, and here there is no choice of values.

First, consider the solid state of potassium, which requires 21.4 kcal (89.5 kJ) per mole to atomize. If we could begin with K_2, it would require only half of the 13.1 kcal (54.8 kJ) per mole dissociation energy, or 6.8 kcal (27.6 kJ) to atomize. The metallic solid is more than three times as stable as the covalent molecule, which accounts for potassium's preference for the solid state, wherein the bonding electrons can be delocalized. Once atomized, potassium presents one outer half-filled orbital per atom, available for forming a single covalent bond. However, consistent with having only one outermost electron, potassium atoms are exceptionally large for the number of 19 electrons, having a nonpolar covalent radius (half the bond length in K_2) of 196.2 pm. In relative compactness these atoms are among the lowest, as the result of having a very small effective nuclear charge operating over a relatively large distance. The electronegativity of potassium is therefore very low, 0.445. Any bonds it might form to other elements would be expected to leave it with a high positive charge, because practically all other elements are more electronegative.

Bromine of course exhibits the liquid state under ordinary conditions, for which the atomization energy is 26.7 kcal (111.7 kJ) per mole of atoms. Since the dissociation energy of the Br_2 molecules in the gaseous state is 46.1 kcal (192.9 kJ) per mole, the atomization energy of the gas is half this, or 23.1 (91.5). The difference between the 26.7 and the 23.1 or 3.6 kcal (15.1 kJ) per mole of atoms represents the strength of the van der Waals forces among Br_2 molecules in the liquid, as measured by the heat of vaporization. This fairly low value (only about one-third that of water) corresponds to the considerable volatility of bromine at ordinary temperatures. Each bromine atom has 35 electrons, with 7 in the outermost shell which can hold 8. Consequently it has the ability to form but one covalent bond. However, the 35 electrons are packed within a sphere of covalent radius 114.2 pm, compared to the potassium atoms where only 19 electrons are held within a covalent radius of 196.2 pm. Thus the bromine atom is much more compact than the potassium atom and would be expected to be more electronegative. With 7 outermost electrons, none of which is more than about one-third effective in blocking off nuclear charge from the others, it is not surprising that a relatively high effective nuclear charge gives the bromine atom the highest electronegativity and the shortest radius of active elements in its period.

Keeping in mind these fundamental qualities of the potassium and bromine atoms, it is easy to predict and understand that when they come together, they will unite one to one to form a KBr molecule in which the single covalent bond is highly polar. From the electronegativities, 0.445 for K and 3.219 for Br, we calculate a geometric mean (the square root of their product) of 1.197. This is the electronegativity of both atoms in the molecule. The electronegativity of potassium increases by 1.197 - 0.445 = 0.752. Acquisition of unit positive charge would have increased the electronegativity (Table 3:1) by 1.047. The partial charge on potassium is therefore 0.752/1.047 = 0.718. The electronegativity of bromine has been diminished by the acquisition of more than half share of the bonding electrons, by 1.197 - 3.219 = -2.022. Acquisition of a complete electron would have reduced the electronegativity by 2.817. The partial charge on bromine in KBr is therefore the ratio -2.022/2.817 = -0.718.

The bond length in a molecule of KBr is measured to be 282.1 pm, compared with a nonpolar covalent radius sum for K and Br of 310.4. The shortening results (see Chapter 15) from the fact that the radial contraction of the atom acquiring partial positive charge exceeds the radial expansion of the atom acquiring partial negative charge. In gaseous KBr, the radii are calculated to be, K 120.0, Br 162.2, for a sum of 282.2pm. The potassium radius contracts by 196.2 - 120.0 = 76.2 pm, whereas the bromine radius expands by only 162.2 - 114.2 = 48.0 pm, for a net bond shortening of 28.2 pm. This is typical, especially where the positive atom is outnumbered by the negative atoms and therefore bears higher charge than any one of them. The geometric mean of the two homonuclear single bond energies, 13.1 and 46.1, is 24.6 kcal (102.9 kJ).

We may now calculate what the bond energy would

be in the KBr molecule if the bond were nonpolar covalent,

$$E_c = 24.6 \times 310.4/282.1 = 27.1 \text{ kcal } (113.4 \text{ kJ})$$

Since it would cost $21.4 + 26.7 = 48.1$ kcal (201.3 kJ) just to prepare the atoms, this is not nearly enough to allow a reaction. This illustrates the importance of **electronegativity difference and bond polarity.**

If on the other hand the bonding electrons were completely controlled by the bromine atom, causing the bond to be completely ionic, the energy would be:

$$E_i = 33200/282.1 = 117.7 \text{ kcal } (492.5 \text{ kJ}) \text{ per mole}$$

The true value for the polar covalent bond that is about 72 per cent ionic can now be obtained as a blend of these extremes, for the actual bond energy will be somewhere between 27.1 and 117.7 kcal per mole.

$$t_c E_c + t_i E_i = 0.282 \times 27.1 + 0.718 \times 117.7$$
$$= 7.6 + 84.5 = 92.1 \text{ kcal } (385.3 \text{ kJ}) \text{ per mole}$$

The best experimental value appears to be 91.5 ± 2 kcal per mole.

It will be shown later (Chapter 14) how this gaseous molecule becomes much more stable by condensing with like molecules to form a crystalline solid.

For reasons not yet understood, calculated bond energies for all the gaseous alkali metal halides agree satisfactorily with the experimental values for only 9 of the 20 compounds, the calculated values being substantially higher for all the others. Since the identical parameters are used in accurate calculation of all 20 of these halides in the solid state (Chapter 14), it seems possible that the experimental values for some of the gaseous molecules, rather than those calculated, may be at fault. However, the deviant values are for all the fluorides and for the other halides of Li and Na, involving the smaller atoms. This suggests a possibility of polarization effects in the deviant molecules which might, in effect, reducing their polarity and thus weakening the bond. Such effects seem large in the alkali metal hydrides, as will be discussed presently. Data for the alkali halide gas molecules are given later in this chapter.

The experimental bond energy for KBr is not known with ideal accuracy but it will serve to illustrate another very important point to be detailed in Chapter 15. The total bond energy equation can be rearranged to a form allowing calculation of t_i:

$$t_i = (E - E_c)/(E_i - E_c)$$

Note that the calculation of the total hypothetically possible E_c does not involve any but experimentally determined homonuclear bond energies and distances, and that the value for E_i depends only on the bond length experimentally determined. There is, therefore, in the above equation for t_i, **no reliance whatever** on **electronegativity, equalization, partial charge,** or **any** of the concepts involved in determining these properties. In other words, if we wish we can view the whole electronegativity-partial charge concept with skepticism, and entirely **independently** calculate t_i:

$$t_i = (91.5 - 27.1)/(117.7 - 27.1) = 64.4/90.6 = 0.711$$

Note how close this is to the 0.718 calculated from partial charges based on electronegativity equalization. Many examples can be found to illustrate this same kind of result (Chapter 15). Whatever may be argued about the validity and significance of partial charges, it cannot be denied that they provide a remarkably satisfactory basis for weighting the relative contributions of covalence and ionicity to the total energy of a polar covalent bond.

Other Examples

Results of bond energy calculations for compounds having only single covalent bonds and using only fully weakened homonuclear bond energies are summarized in Tables 3:2-3:6. These include compounds of positive hydrogen, about which more need be said. Hydrogen is unique in a number of ways. One is that although half the bond length in H_2 is 37 pm, the nonpolar covalent radius of H in its compounds appears better represented as 32 pm. Another is that for hydrogen alone the homonuclear covalent bond energy must

Table 3:2

Bond Energies in Binary Hydrogen Compounds Having Only Single Bonds

cmpd	bond	R_o, pm	t_i	E_c	E_i	Ecalc	Eexp
HF	H–F'	91.8	0.248	46.2	89.7	135.9	135.9
HCl	H–Cl'	127.4	0.162	61.5	42.2	103.7	103.2
HBr	H–Br'	140.8	0.117	59.7	27.6	87.3	87.4
HI	H–I'	160.8	0.036	59.7	7.4	67.1	71.4
H_2O	H–O'	95.8	0.187	48.0	64.8	112.8	110.8
H_2O_2	H–O'	95.0	0.192	46.2	67.1	113.3	
	O'–O'	147.5	0.0	32.0	0.0	32.0	
						258.6	255.9
H_2S	H–S'	133.	0.069	70.8	17.2	88.0	87.7
H_2S_2	H–S'	133.	0.070	69.8	17.5	87.3	
	S'–S'	205.	0.0	56.2	0.0	56.2	
						230.7	232.9
H_2Se	H–Se'	146.0	0.079	57.6	18.0	75.6	75.7
H_2Te	H–Te'	165.3	0.005	62.6	1.0	63.6	63.7
NH_3	H–N'	101.2	0.110	57.3	36.1	93.4	93.4
N_2H_4	N'–N'	143.7	0.0	39.6	0.0	39.6	
						416.9	411.6
PH_3	H–P'	143.7	0.016	73.1	3.7	76.8	76.7
P_2H_4	P'–H	142.	0.015	74.0	3.5	77.5	
	P'–P'	212.	0.0	56.1	0.0	56.1	
						366.1	362.2
AsH_3	H–As'	151.9	0.043	61.7	9.4	71.1	70.9
SbH_3	Sb'–H	170.7	0.022	57.3	4.3	61.6	61.6
CH_4	H–C	109.4	0.030	90.0	9.1	99.1	99.4
GeH_4	Ge–H	152.7	0.003	68.9	0.7	69.6	69.1
Ge_2H_6	Ge–H	154.	0.005	69.2	1.1		
	Ge–Ge	241.	0.0	45.5	0.0	45.5	
						460.7	452.8
Ge_3H_8	Ge–H	154.	0.005	69.2	1.1	70.3	
	Ge–Ge	241.	0.0	45.5	0.0	644.6	631.1
NH_2OH	H–N'	103.	0.114	54.0	36.7	90.7	
	N'–O'	146.	0.073	32.9	16.6	49.5	
	H–O'	97.	0.187	47.5	64.0	111.5	
						342.4	341.9

be corrected for the positive charge, for without this correction all calculated values are significantly too high. The homonuclear energy, 104.2 kcal (436.0 kJ) per mole, is multiplied by the factor, 1.000 – d_H. As evident in the data of Table 3:2, this correction allows the accurate calculation of bond energies in compounds of positive hydrogen. Finally, wherever the hydrogen bears partial **negative** charge, the

Table 3:3

Bond Energies in Gaseous Molecules of F' Compounds

cmpd	bond	R_o, pm	t_i	E_c	E_i	Ecalc	Eexp
LiF	Li-F'	156.4	0.753	10.1	160.3	170.4	138.
NaF	Na-F'	192.6	0.797	6.0	137.1	143.1	114.5
KF	K-F'	217.	0.851	4.2	130.2	134.4	119.
RbF	Rb-F'	227.	0.920	2.2	134.6	136.8	124.
CsF	Cs-F'	234.5	0.977	0.6	138.3	138.9	127.
MgF_2	Mg-F'	177.	0.591	19.6	111.0	130.6	131.3
CuF	Cu-F'	174.3	0.366	26.1	69.8	95.9	103.
AgF	Ag-F'	196.	0.413	22.2	70.0	92.2	84.5
ZnF_2	Zn-F'	181.	0.345	27.2	63.3	90.5	92.0
CdF_2	Cd-F'	197.	0.400	23.2	67.4	90.6	
HgF_2	Hg-F'	200.	0.347	13.3	57.6	70.9	
PbF_4	Pb-F'	208.	0.338	21.5	54.0	75.5	77.1
NF_3	N-F'	137.	0.140	35.2	33.9	69.1	66.7
N_2F_4	N'-F'	137.	0.138	35.3	33.4	68.7	
	N'-N'	147.	0.0	38.7	0.0	38.7	
						313.5	303.
OF_2	O'-F'	141.8	0.057	33.9	13.3	47.2	44.9
SF_4	S'-F'	156.	0.187	42.6	39.8	82.4	78.6
SF_6	S'-F'	156.	0.188	42.5	40.0	82.5	78.6
SeF_6	Se'-F'	168.	0.177	35.6	35.0	70.6	72.5
TeF_6	Te'-F'	184.	0.262	31.8	47.3	79.1	79.2
ClF	Cl'-F'	163.	0.087	45.5	17.7	63.2	59.9
BrF	Br'-F'	176.	0.131	38.9	24.7	63.6	59.6
IF	I'-F'	191.	0.212	31.7	36.9	68.6	67.0
HOF	H-O'	96.	0.201	42.0	69.5	111.5	
	O'-F'	140.	0.053	34.5	12.6	47.1	
						158.6	156.7
ClO_3F	Cl'-F'	163.	0.086	45.6	17.5	63.1	
	Cl'-O'	142.	0.031	50.4	7.2	57.6	
						235.9	232.3

calculated bond energy is too large. For highly polar hydrides, the calculated energies are far too high, the experimental energy being much closer to that which would be expected if the bond were nonpolar. A quantitative theory of bonding to negative hydrogen has not yet been developed. It seems likely that the exceptionally high polarizability of hydride allows distortion of the electron sharing such as practically to nullify the energy effects expected of the polarity.

Fully weakened single covalent bonds are found not

Table 3:4

Bond Energies in Gaseous Molecules of Cl' Compounds

cmpd	bond	R_o, pm	t_i	E_c	E_i	Ecalc	Eexp
LiCl	Li-Cl'	202.1	0.666	14.6	109.5	124.1	112.
NaCl	Na-Cl'	236.1	0.711	9.6	100.0	109.6	98.
KCl	K-Cl'	266.7	0.764	7.2	95.1	102.3	101.7
RbCl	Rb-Cl'	279.	0.833	5.1	99.2	104.3	107.
CsCl	Cs-Cl'	290.6	0.890	3.2	101.7	104.9	105.
$BeCl_2$	Be-Cl'	177.	0.349	43.3	65.5	108.8	110.6
$MgCl_2$	Mg-Cl'	218.	0.493	27.2	75.1	102.3	103.4
CuCl	Cu-Cl'	205.1	0.279	34.6	45.2	79.8	91.5
AgCl	Ag-Cl'	228.1	0.327	29.9	47.6	77.5	76.
$ZnCl_2$	Zn-Cl'	205.	0.248	38.3	40.2	78.5	78.7
$CdCl_2$	Cd-Cl'	221.	0.305	32.8	45.8	78.6	
$HgCl_2$	HgCl'	234.	0.253	17.8	35.9	53.7	53.9
$PbCl_4$	Pb-Cl'	243.	0.238	29.1	32.5	61.6	59.4
PCl_3	P-Cl'	204.	0.184	46.9	29.9	76.8	77.1
Cl_2O	Cl'-O'	170.	0.030	42.7	5.9	48.6	49.3
SCl_2	S'-Cl'	200.	0.093	52.3	15.4	67.7	64.8
$SeCl_2$	Se'-Cl'	213.	0.083	43.8	12.9	56.7	60.0
BrCl	Br'-Cl'	213.6	0.045	49.3	7.0	56.3	52.3
ICl	I'-Cl'	232.1	0.126	40.1	18.0	58.1	50.5
HOCl	H-O'	96.	0.196	44.1	67.8	111.9	
	Cl'-O'	172.	0.029	42.2	5.6	47.8	
						159.7	162.7
SiH_3Cl	Si-H	148.	o.100	67.3	22.4	89.7	
	Si-Cl'	205.	0.253	44.1	41.0	85.1	
						354.2	341.3
SiH_2Cl_2	Si-H	148.	0.103	64.9	23.1	88.0	
	Si-Cl'	205.	0.260	43.7	42.1	85.8	
						347.6	345.2

only in the 98 compounds of Tables 3:2–3:6, but also in hundreds of other molecules along with less weakened or multiple bonds. Among these are all hydrocarbons and their derivatives.

In applying this theory of polar covalence to the quantitative evaluation of contributing bond energies, it is important to be mindful of two possible influences, protonic bridging and steric interference. If the molecular structure includes positive hydrogen and a small negatively charged atom having a lone pair of electrons, so positioned as to permit internal protonic bridging, it may occur, adding 2–8 kcal to the

Table 3:5

Bond Energies in Gaseous Molecules of Br' Compounds

cmpd	bond	R_o, pm	t_i	E_c	E_i	Ecalc	Eexp
LiBr	Li-Br'	217.	0.621	14.6	95.0	109.6	101.
NaBr	Na-Br'	250.2	0.666	9.8	88.4	98.2	88.5
KBr	K-Br'	282.1	0.719	7.6	84.6	92.2	91.5
RbBr	Rb-Br'	294.5	0.789	5.7	88.9	94.6	93.
CsBr	Cs-Br'	307.2	0.846	3.9	91.4	95.3	95.
$BeBr_2$	Be-Br'	191.	0.300	41.2	52.1	93.3	92.9
$MgBr_2$	Mg-Br'	234.	0.442	26.3	62.7	89.0	90.3
CuBr	Cu-Br'	214.	0.235	33.6	36.5	70.1	79.
AgBr	Ag-Br'	239.2	0.282	28.7	39.1	67.8	69.
$ZnBr_2$	Zn-Br'	221.	0.202	35.7	30.3	66.0	66.0
$CdBr_2$	Cd-Br'	237.	0.257	30.9	36.0	66.9	
$HgBr_2$	Hg-Br'	244.	0.205	17.1	27.9	45.0	44.3
$PbBr_4$	Pb-Br'	258.	0.188	27.6	24.2	51.8	49.4
PBr_3	P-Br'	220.	0.137	43.9	20.7	64.6	63.4
IBr	I'-Br'	248.5	0.081	37.3	11.0	48.3	42.8
$SiHBr_3$	Si-H	146.	0.104	65.2	23.6	88.8	
	Si-Br'	216.	0.221	41.6	34.0	75.6	
						315.6	316.2

experimental atomization energy which is not included in the calculated value. As a possible example, lack of freedom of rotation around the HO-N bond in HNO_3 has been interpreted as evidence of an internal protonic bridge. This may account for the difference between calculated (374.1 kcal (1565 kJ)) and experimental (376.1 kcal (1574 kJ) atomization energies. If atoms or groups of atoms are so positioned as to crowd one another, bonding may be weakened and calculated atomization energies will be larger than experimental values. For example, the calculated atomization energy of CCl_4 is 346.4 kcal (1449 kJ), compared to only 312.5 kcal (1308 kJ) per mole for the experimental value. This seems to be the result of crowding the four chlorine atoms around the smaller carbon atom.

In Figure 3:2 are summarized the bond energies of gaseous fluorides so that the relative contributions of covalence and ionicity can easily be seen. Figure 3:3 provides similar information about gaseous binary chlorides, except that here only the percentage contributions are shown. Both figures

Figure 3:2

Bond Energies in Gaseous Binary Fluorides
(white–covalent; black–ionic)

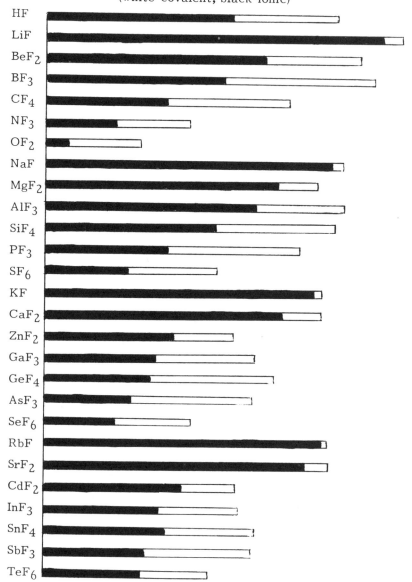

Figure 3:3

Percentage of Covalent and Ionic Energy
in Gaseous Chlorides

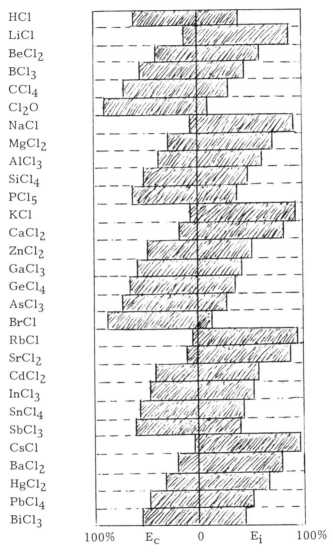

HCl
LiCl
BeCl$_2$
BCl$_3$
CCl$_4$
Cl$_2$O
NaCl
MgCl$_2$
AlCl$_3$
SiCl$_4$
PCl$_5$
KCl
CaCl$_2$
ZnCl$_2$
GaCl$_3$
GeCl$_4$
AsCl$_3$
BrCl
RbCl
SrCl$_2$
CdCl$_2$
InCl$_3$
SnCl$_4$
SbCl$_3$
CsCl
BaCl$_2$
HgCl$_2$
PbCl$_4$
BiCl$_3$

100%   E$_c$ 0   E$_i$ 100%

Table 3:6

Bond Energies in Molecules of I' Compounds

cmpd	bond	R_o, pm	t_i	E_c	E_i	Ecalc	Eexp
LiI	Li-I'	239.	0.540	15.3	74.9	90.2	84.
NaI	Na-I'	271.1	0.585	10.7	71.7	82.4	72.
KI	K-I'	304.8	0.638	8.5	69.5	78.0	79.
RbI	Rb-I'	317.7	0.711	6.7	74.3	81.0	80.
CsI	Cs-I'	332.	0.764	5.2	76.5	81.7	81.
BeI$_2$	Be-I'	212.	0.212	40.8	33.2	74.0	71.9
MgI$_2$	Mg-I'	257.	0.351	26.6	45.3	71.9	71.2
CuI	Cu-I'	233.4	0.153	33.0	22.1	54.3	47.
AgI	Ag-I'	254.4	0.201	28.5	26.3	54.7	56.
ZnI$_2$	Zn-I'	238.	0.117	35.0	16.3	51.3	50.1
CdI$_2$	Cd-I'	255.	0.170	30.4	22.1	52.5	
HgI$_2$	Hg-I'	259.	0.120	16.9	15.4	32.3	34.6
BI$_3$	B-I'	203.	0.104	50.0	17.0	67.0	66.2
PbI$_4$	Pb-I'	277.	0.100	27.1	12.0	39.1	37.3
PI$_3$	P-I'	247.	0.052	41.2	7.0	48.2	
$(C_4H_9)_2BI$							
	B-I'	203.	0.103	50.1	16.8	66.9	
	B-C	156.	0.098	74.7	20.9	95.6	
	C_4H_9					1137.6	
						2533.2	2526.6

reveal very clearly the periodicity that exists in the nature of the bonds.

The simple theory of polar covalence is applicable without modification to many compounds, but of course, not all. As Professor Blair Saxton at Yale used to tell us, "It is we who are simple, not Nature." The truth of this recognition must impress us all increasingly as our experience expands. However, with appropriate modification, the theory can be successfully applied to nearly all major group compounds capable of existence in gaseous molecular form, and to many solids as well. Two factors needing further discussion and explanation at this point are bond multiplicity and lone pair bond weakening. These are the subjects of the next two chapters.

Chapter 4

MULTIPLE BONDS

MULTIPLICITY FACTORS

The bond energy of multiple bonds can be determined by calculating the energy in the usual way as though the bonds were single but of the experimental length. Then the entire energy, both covalent and ionic contributions, is multiplied by an appropriate factor. As mentioned in Chapter 2, the double bond factor is 1.488, derived from a study of alkenes but applicable to O_2, S_2, C=O, etc. bonds everywhere. There are very few triple bonds other than in alkynes, for which the factor is 1.787, but this factor appears applicable to all. From the experimental atomization energy of benzene and the assumption that the standard C-H energy of 98.81 kcal (413.4 kJ) applies, it is possible to determine a multiplicity factor for the ring CC bonds of 1.299. These multiplicity factors are a linear function of the cube root of the bond order (which is half the total number of electrons in the bond):

$$f = 1.640(BO)^{1/3} - 0.578$$

Multiplicity in the bonding is always associated with a decrease in the lone pair bond weakening effect. Triple bonds remove it completely and double bonds halfway. However, in both diatomic and polyatomic gaseous molecules, multiple bonding appears to be inhibited by disparity in atomic radius and by angles between orbitals of 120° or greater. Of 86 molecules which would be expected to contain normal double or triple bonds on the basis of the compound formula and the normal valences of their atoms, only about 36 do. The remainder exhibit bonding of lower than expected order, as determined by comparison of the experimental bond energies with those calculated by the theory of polar covalence.

GASEOUS DIATOMIC OXIDES

Relevant data on gaseous oxides are most available. In addition to the thousands of familiar compounds containing oxygen, there are a number of diatomic oxides that have been studied in the gaseous state in sufficient detail that certain observations can be made. Many of these oxides do not satisfy the normal bonding capacity of both atoms. The theory of polar covalence provides a means of recognizing the bond types from the experimental bond energies and bond lengths, provided the normal valences are exhibited. In Chapter 13 it will be shown that in subvalent molecules, wherein the full bonding capacity of one or more atoms is not used, bond energies are significantly affected. When this results in an odd number of nonbonding electrons in the outermost principal quantum level, the bonds appear somewhat weaker than in molecules wherein normal valences are observed. When it leaves an even number of nonbonding electrons in the outermost level, the bonds appear somewhat stronger than expected. However, these weakening and strengthening effects do not appear to affect multiple bonding to the degree that they affect single bonds. We may therefore apply the theory of polar covalence to multiple bonding, with due caution, even where the normal valences are not shown. There is no evidence in this work that the parameters of Table 3:1 are not accurately applicable to different oxidation states, for example, the values for sulfur being equally applicable to H_2S and H_2SO_4.

The double bond factor 1.488 given earlier was derived from period 2 elements and its applicability to heavier elements might be suspect. However, the same factor applies to S_2, Se_2, and Te_2. Similarly, the same triple bond factor, 1.787, is applicable to N_2, P_2, As_2, and Sb_2. Therefore these factors may be generally applied with reasonable confidence.

In the gaseous diatomic molecules of M2 oxides, Table 4:1, where the stoichiometry would suggest normal valence, the bond dissociation energies reveal that the bonds cannot be double as might be expected, but are single. This is probably the consequence of the opposite orientation of the bonding orbitals of the M2 atom, which appears to prevent formation of a double bond to oxygen and leaves the diatomic molecule in effect a diradical. This corresponds to one free electron on each atom and results in an actual bond energy significantly lower than calculated for a normal single polar covalent bond at the observed distance.

Table 4:1

Bond Energies in Gaseous Monoxides

cmpd	bond	R_o, pm	t_i	E_C	E_i	Ecalc	Eexp
BeO	Be-O'	133.1	0.361	36.4	90.1	126.5	107.
MgO	Mg-O'	174.9	0.486	23.0	92.3	115.3	104.
CaO	Ca-O'	182.2	0.598	17.0	106.8	123.8	111.
SrO	Sr-O'	192.0	0.677	12.6	117.1	129.7	109.
BaO	Ba-O'	194.0	0.704	11.2	120.5	131.7	135.
BO	B=O'''	120.	0.257	125.4	105.8	231.2	
	B-O'''	120.5	0.257	84.3	71.1	155.4	
					aver	193.3	192.8
AlO	Al-O''	161.8	0.384	42.9	78.7	121.6	122.
GaO	Ga-O'	174.	0.227	32.0	43.3	75.3	68.
InO	In-O''	199.	0.286	35.2	47.7	82.9	86.
CO	C≡O'''	112.8	0.162	184.3	85.2	269.5	257.3
SiO	Si=O'''	150.9	0.286	98.8	93.6	192.4	191.
GeO	Ge=O'''	165.1	0.186	98.2	56.7	154.9	158.
SnO	Sn=O''	183.8	0.252	63.4	67.4	130.8	131.
PbO	Pb=O''	192.2	0.253	51.5	65.0	116.5	
	Pb-O'	192.2	0.253	24.2	43.7	67.0	
					aver	92.2	90.
NO	N''=O''	115.0	0.079	116.0	33.9	149.9	150.7
PO	P''=O''	147.4	0.207	90.9	69.6	160.5	
	P'-O'''	147.4	0.207	72.9	46.8	119.7	
					aver	140.1	142.6
AsO	As''=O''	157.	0.149	85.2	46.9	132.1	
	As''-O'''	157.	0.149	67.0	31.5	98.5	
					aver	115.3	115.
SbO	Sb''=O''	174.	0.219	68.6	62.2	130.8	
	Sb''-O'''	174.	0.219	57.3	41.8	99.1	
					aver	114.9	102.
BiO	Bi-O''	193.	0.243	40.1	41.8	81.9	81.9
O₂	O''=O''	120.7	0.0	119.1	0.0	119.1	119.1
SO	S''=O''	149.3	0.122	98.8	40.4	139.2	
	S'-O'''	149.3	0.122	77.8	27.1	104.9	
					aver	122.1	124.7
SeO	Se''=O''	164.	o.112	84.8	33.7	118.6	
	Se''-O'''	164.	0.112	70.0	22.7	92.7	
					aver	105.6	101.
TeO	Te''=O''	183.	0.187	70.7	50.5	121.2	
	Te'-O'	183.	0.187	32.4	33.9	66.3	
					aver	93.8	93.
ClO	Cl'-O'	155.	0.030	46.8	6.4	53.2	
	Cl'-O''	155.	0.030	67.0	6.4	73.4	
					aver	63.3	65.
BrO	Br'-O'	167.	0.077	40.2	15.3	55.5	56.2

In gaseous diatomic oxides of M3, even boron, which in its normal trivalence readily forms normal double bonds to oxygen, as in BOF, BOCl, and $B_2O_3(g)$, appears incapable of true double bonding and instead forms a bond better described as of order 1.5. In AlO, GaO, and InO there appear to be only single bonds. Even in the gaseous oxyhalides, aluminum appears incapable of forming a normal double bond to oxygen, for the atomization energies of AlOF and AlOCl appear to require the weaker bonding to oxygen associated with a bond order of 1.5. Here is our first example of the effect of disparity in atomic size. The data supporting these and subsequent statements are presented in Tables 4:1 and 4:3.

With M4 elements, oxygen can form diatomic molecules in which a triple bond is theoretically possible through transfer of an electron of oxygen to pair with a bonding electron of the other atom. However, only carbon seems to form such a bond, the bond orders in the other gaseous monoxides of this group being: SiO 2, GeO 2, SnO 2 and PbO 1.5. When oxygen is attached to an M4 atom exhibiting its normal valence, the expected double bond is seen only with carbon. Almost always oxygen in a carbonyl group is attached by a normal double bond, as in CO_2 and all aldehydes, ketones, carboxylic acids, amides, and esters. The only exceptions discovered to date are the carbonyl halides, COFCl and $COCl_2$ (but not COF_2), wherein, for reasons not yet understood, the CO bond appears to be of order 1.5 rather than the expected 2. That is, the bond can be represented quantitatively as an average of an ordinary double bond and a single bond in which oxygen is accepting a pair of electrons from the other atom into an orbital made available through the pairing of its normal bonding electrons: C=O", C-O'".

Similar bonds appear to occur in gaseous SiO_2, where the bond order appears also to be 1.5. Data are not available for other gaseous dioxides of this group, but a bond order of 1.5 or lower is easily predictable, as the size disparity increases.

In nitric oxide, NO, the bond order appears to be clearly 2, not 2.5 as usually claimed. In the other M5 monoxide molecules, however, the bond is less than double, the bond order being 1.5 in PO, AsO, and SbO, and only 1 in BiO.

Table 4:2

Bond Energies in Gaseous Chalcides

cmpd	bond	R_o, pm	t_i	E_c	E_i	Ecalc	Eexp
BS	B=S''	160.9	0.135	101.9	41.4	143.3	139.
B_2S_3	B–S'''	180.	0.137	63.8	25.3	89.1	
	B=S'''	165.	0.137	103.5	41.0	144.5	
						467.2	487.6
AlS	Al–S''	203.	0.261	45.4	42.7	88.1	89.
CS	C≡S'''	153.	0.040	153.1	15.5	168.6	167.
SiS	Si≡S'''	192.9	0.164	105.4	42.0	147.4	148.
GeS	Ge≡S'''	201.	0.065	102.8	19.1	121.9	131.5
SnS	Sn≡S'''	221.	0.130	84.4	34.9	119.3	
	Sn=S'''	221.	0.130	70.3	29.1	99.1	
					aver	109.3	111.
PbS	Pb=S''	229.5	0.131	54.4	28.2	82.6	82.7
SN	S'–N'''	149.6	0.169	71.8	37.7	109.5	111.
PS	P''=S''	192.	0.085	89.6	21.9	111.5	106.
CS_2	C=S'''	156.	0.040	125.1	12.7	137.8	137.8
S_2	S''=S''	188.7	0.0	100.0	0.0	100.0	101.6
Se_2	Se''=Se''	215.	0.0	74.3	0.0	74.3	79.5
CSe	C≡Se'''	168.	0.050	110.6	14.7	125.3	
	C=Se'''	168.	0.050	132.8	17.7	150.5	
					aver	137.9	141.
SiSe	Si=Se''	206.	0.174	69.5	41.7	111.2	
	Si≡Se'''	206.	0.174	90.2	50.1	140.3	
					aver	125.8	127.
GeSe	Ge≡Se'''	213.	0.075	91.1	20.9	112.0	117.
SnSe	Sn=Se'''	233.	0.140	62.4	28.9	91.3	95.9
PbSe	Pb=Se'''	240.	0.142	47.0	29.2	76.2	72.4
Te_2	Te''=Te''	259.	0.0	63.2	0.0	63.2	63.2
GeTe	Ge≡Te'''	234.	0.0	84.9	0.0	84.9	96.
SnTe	Sn=Te'''	252.	0.065	58.9	12.7	71.6	76.3
PbTe	Pb≡Te'''	260.	0.067	46.7	12.7	59.4	60.

In M6, oxygen of course forms O_2 molecules in which the bond order is clearly 2, but in both SO and SO_2, as well as SeO and TeO, it is 1.5.

SULFIDES

Sulfur appears somewhat more adaptable than oxygen, in being able to form true triple bonds not only to silicon in its own period, but also to smaller C and larger Ge (T.4:2).

However, the size discrepancy appears too great in SnS and PbS, SnS apparently having bond order 2.5 and PbS, 2. For diatomic selenides, the bond orders are CSe 2, SiSe 2.5, GeSe 3, and SnSe and PbSe 2.

OTHER COMPOUNDS

Data for other compounds having possible multiple bonds are presented in Table 4:3. Although bond dissociation energies are known at least approximately for many other diatomic gas molecules, absence of reliable bond lengths or a method of estimating such lengths prevents more than a highly speculative study of the nature of multiple bonding within them. For the present it may be stated that in general the factors governing the degree of multiplicity appear to be similar to those discussed herein.

THE BURNING OF CARBON

The most important chemical reaction in the world today, from the viewpoint of obtaining useful energy for man's needs, is the combination of carbon with oxygen. The theory of polar covalence permits unique insights into the nature of this reaction, in terms of the properties of carbon and oxygen atoms and the origin of the energy which is obtained. A thorough discussion is therefore justified here.

The first step toward the condensation of atomic carbon is the formation of C_2 molecule. This molecule has a bond energy, 146.5 kcal (613.0 kJ) per mole, and a bond length of 131 pm, both of which correspond to a double bond. Formation of a double bond between two carbon atoms is known to direct the other two bonding orbitals of each carbon atom approximately to the same planar structure as that of ethylene, the bond angles being about 120°. (Curiously, addition of one hydrogen atom to each carbon directs the remaining fourth orbital to the formation of a triple bond between the two carbon atoms, in acetylene.) The C_2 molecule is therefore a tetraradical, and its further condensation is bound to result in the planar condensed ring system of graphite. This is probably the explanation for the scarcity of diamond as compared with graphite. At any rate, the energy of the three bonds per carbon of 1.33 order together with the van der Waals attractions between layers leads to an atomization

Table 4:3

Bond Energies in Other Compounds Having Possible Multiple Bonds

cmpd	bond	R_o, pm	t_i	E_c	E_i	Ecalc	Eexp
B_2	B–B	159.	0.0	79.3	0.0	70.3	71.
B_2O_3	B=O"	124.	0.263	97.9	104.8	202.7	
	B–O"	136.	0.263	60.0	64.2	124.2	
						653.8	648.1
BOF	B–F"'	130.	0.324	72.8	82.7	155.5	
	B=O"	124.	0.256	98.8	102.0	200.8	
						356.3	357.3
BOCl	B–Cl'	175.	0.237	52.8	45.0	97.8	
	B=O"	124.	0.265	97.6	105.6	203.2	
						301.0	297.4
BOBr	B–Br'	187.	0.210	46.5	37.2	83.7	
	B=O"	124.	0.262	98.0	104.4	202.4	
						286.1	280.1
BN	B–N'	128.	0.178	54.6	46.2	100.8	93.
HOBO	H–O'	96.	0.189	47.1	65.4	112.5	
	B–O"	136.	0.261	60.1	63.7	123.8	
	B=O"	124.	0.261	98.1	104.0	202.1	
						438.4	439.4
AlN	Al–N'	168.	0.305	35.6	60.3	95.9	94.
AlOF	Al–F"	165.	0.466	38.2	93.8	132.0	
	Al=O"	162.	0.417	59.8	127.2	187.0	
	Al–O"'	162.	0.417	49.4	85.5	134.9	
						293.0	296.3
AlOCl	Al–Cl'	214.	0.379	34.6	58.8	93.4	
	Al=O"	162.	0.406	59.8	123.8	183.6	
	Al–O"'	162.	0.406	49.7	83.2	132.9	
						251.7	250.6
C_2	C=C	131.	0.0	149.8	0.0	149.8	145.
CN	C–N"'	117.	0.083	105.9	23.6	129.5	
	C≡N"'	117.	0.083	189.2	42.2	231.4	
					aver	180.5	184.
CNF	C–F',F"	126.	0.217	63.1	57.2	120.3	
	C=N"'	116.	0.087	158.8	36.1	194.9	
						315.2	320.2
CNCl	C–Cl'	162.9	0.133	66.2	27.1	93.3	
	C=N"'	116.	0.085	159.1	36.2	195.3	
						288.6	280.3
CNBr	C–Br'	179.0	0.089	61.1	16.5	77.6	
	C=N"'	115.7	0.084	159.7	35.9	195.6	
						273.2	266.5
CNI	C–I'	199.5	0.007	58.2	1.2	59.4	
	C=N"'	115.9	0.082	159.7	35.0	194.7	
						254.1	257.4

POLAR COVALENCE

Table 4:3 cont.

cmpd	bond	R_o, pm	t_i	E_c	E_i	Ecalc	Eexp
CO_2	C=O"	116.	0.166	120.9	70.7	191.6	192.3
COS	C=O"	116.	0.161	124.7	68.6	190.3	
	C=S'''	156.	0.042	124.8	13.3	138.1	
						328.4	331.1
COF_2	C-F"	131.	0.227	69.4	57.5	126.9	
	C=O"	117.	0.172	117.4	72.6	190.0	
						422.4	420.4
COFCl	C-F',F"	132.	0.223	59.9	56.1	116.0	
	C-Cl'	175.	0.140	61.2	26.6	87.8	
	C=O"	116.	0.169	120.6	72.0	192.6	
	C-O'''	116.	0.169	99.5	48.4	147.9	
						374.0	379.8
$COCl_2$	C-Cl'	175.	0.137	61.3	26.0	87.3	
	C=O"	117.	0.166	119.5	70.1	189.6	
	C-O'''	117.	0.166	99.0	47.1	146.1	
						342.4	341.2
Si_2	Si=Si	225.	0.0	83.6	0.0	83.6	78.1
SiN	Si-N'''	157.	0.207	68.9	43.8	112.7	105.
SiO_2	Si=O"	154.	0.300	77.2	96.2	173.4	
	Si-O'''	154.	0.300	63.8	64.7	128.5	
					aver	150.9	151.7
Ge_2	Ge=Ge	227.	0.0	72.0	0.0	72.0	65.
N_2	N'''≡N'''	109.8	0.0	226.0	0.0	226.0	225.9
N_2O	N'''-N'''	113.	0.0	123.3	0.0	123.3	
	N"=O"	118.8	0.078	112.4	32.4	144.8	
						268.1	266.1
NO_2	N'-O"	119.2	0.080	57.3	22.3	79.6	
	N"=O"	119.2	0.080	111.8	33.2	145.0	
					aver	112.3	112.1
NO_2F	N'-F'	140.	0.137	34.5	32.5	67.0	
	N'-O'	121.	0.082	39.3	24.5	61.8	
	N"=O"	121.	0.082	109.8	33.5	143.3	
						272.1	277.0
NO_2Cl	N'-Cl'	183.	0.050	42.5	9.1	51.6	
	N'-O'	121.	0.080	39.4	22.0	61.4	
	N"=O"	121.	0.080	110.2	32.7	142.9	
						255.9	258.2
HONO	H-O'	96.	0.197	43.6	68.1	111.7	
	N'-OH	146.	0.077	32.8	17.5	50.3	
	N"=O"	120.	0.077	111.4	31.7	143.1	
						305.1	303.1
$HONO_2$	H-O'	96.	0.200	42.5	69.2	111.7	
	N'-O'H	·136.	0.078	35.1	19.0	54.1	
	N'-O',O"	123.	0.078	47.3	21.1	68.4	
	N"=O"	123.	0.078	108.6	31.3	139.9	
						374.1	376.1

Table 4:3 cont. 2

cmpd	bond	R_o, pm	t_i	E_c	E_i	Ecalc	Eexp
PN	P"=N"	149.	0.137	97.9	45.4	143.3	147.
P_2	P'''≡P'''	189.	0.0	125.6	0.0	125.6	125.1
As_2	As'''≡As'''	217.	0.0	97.0	0.0	97.0	91.
S_2O	S'''=S'''	188.4	0.0	111.5	0.0	111.5	
	S'-O'''	146.5	0.120	79.5	27.2	106.7	
						218.2	217.6
SO_2	S"=O"	143.2	0.126	102.6	43.5	146.1	
	S'-O'''	143.2	0.126	80.8	29.2	110.0	
					aver	128.1	128.1
SO_3	S"=O"	143.	0.126	107.0	43.5	150.5	
	2S'-O"	143.	0.126	65.8	29.3	95.1	
					aver	113.6	113.2

energy of 171.3 kcal (716.7 kJ) per mole of carbon. The presence of four outermost electrons per atom where 8 are permitted allows each atom to form four single covalent bonds with no vacancies or electrons left uninvolved. It also ensures an intermediate nature of carbon atoms, with intermediate effective nuclear charge and radius, and therefore electronegativity. The electronegativity is 2.746, the radius 77.2 pm, and the homonuclear single covalent bond energy (except in hydrocarbons and derivatives as described in Chapter 6) is that of diamond, 85.4 kcal (357.3 kJ) per mole of bonds.

The reasons for the existence of oxygen as gaseous O_2 have previously been discussed. Each atom of oxygen has 6 outermost electrons where there could be 8, and consequently can provide two half-filled orbitals for bonding. This corresponds to a high effective nuclear charge, exceeded only by that in fluorine, and a small radius of 70.2 pm, leading to the second highest of all electronegativities, 3.654. The bond energy in O_2 is 119.1 kcal (498.3 kJ) per mole.

It is clear that one carbon atom can accommodate two oxygen atoms and that these oxygen atoms must break loose from the molecule before they can unite with carbon. Therefore considerable energy must be expended in preparing the atoms for combining, 171.3 kcal for the carbon and 119.1 for the two oxygens, a total cost of 290.4 kcal (1215 kJ) per mole of CO_2 to be formed. Unless the combination of carbon with oxygen produces more than this amount of energy, no useful energy will be released. We need now to calculate what energy will result from the burning of carbon.

The electronegativity of CO_2 must be the cube root of the product of the electronegativities of all three atoms, which is 3.322. The carbon has changed by 3.322 - 2.746 = 0.576, but acquisition of unit positive charge would have changed it by 2.602. The partial charge on carbon is therefore 0.576/2.602 = 0.221. Similarly, the oxygen has changed by 3.322 - 3.654 = -0.332, whereas acquisition of unit negative charge would have changed it by 3.001. The partial charge on oxygen is therefore -0.322/3.001 = -0.111. The ionic blending coefficient is (0.221 + 0.111)/2 = 0.166, making the covalent blending coefficient 1.000 - 0.166 = 0.834. The bond length is 116.0 pm and the covalent radius sum is 147.4 pm. Taking 85.4 for the homonuclear bond energy of carbon and 68.8 for the E" value for oxygen, appropriate for double bonding, we obtain a geometric mean of 76.7 kcal (321.3 kJ) per mole. The double bond factor is 1.488.

$$E = 0.834 \times 76.7 \times 147.4 \times 1.488/116.0$$
$$+ 0.166 \times 33200 \times 1.488/116.0$$
$$= 120.9 + 70.7 = 191.6 \text{ kcal (801.7 kJ) per mole of bonds}$$

The experimental value is 192.3 kcal (804.6 kJ) per mole of bonds, the difference being 0.4 per cent.

Since 2 x 192.3 or 384.6 kcal is recovered from the combination of the carbon and oxygen atoms, and it cost only 290.4 kcal to liberate these atoms, the difference, 94.2 kcal (394.1 kJ) per mole, is the useful energy, the heat of combustion of carbon.

But why does carbon dioxide exist as CO_2 molecules? Shouldn't it be more stable as a polymeric solid, in which four single bonds would replace the two double bonds? If single bonds were formed, the homonuclear bond energy of oxygen would then be the fully weakened value, 33.6, leading to a geometric mean of 53.6 kcal (224.2 kJ). The bond length would be about 143 pm. Then

$$E = 0.834 \times 53.6 \times 147.4/143$$
$$+ 0.166 \times 33200 /143$$
$$= 45.9 + 38.5 = 84.4 \text{ kcal (353.1 kJ) per mole of bonds}$$

Four such bonds would be 337.8 kcal (1413 kJ) per mole of CO_2. The polymer would be 46.8 kcal (195.8 kJ) per mole less stable than the monomer, and burning carbon would provide that much less heat if the product were the polymer. The polymer would undoubtedly depolymerize spontaneously, the change being favored by an increase both in entropy in forming the gas and in bond strength in forming the double bonds. It is the reduction in lone pair bond weakening that occurs in the oxygen which causes double bonds to be favored over single bonds in this situation.

Another question worth asking is, what if carbon burned to CO_2 molecules which were nonpolar in bonding? The bond length would be greater, perhaps 125 pm instead of 116 pm, and the bond would be entirely covalent:

$$E = 76.7 \times 1.488 \times 147.4/125 = 134.2 \text{ kcal (561.6 kJ) per mole}$$

Two moles of bonds would release 268.4 kcal, not even enough to pay the cost (290.4) of atomizing the carbon and oxygen. If the electronegativity were the **same** for carbon and oxygen, **carbon would not burn.**

In conclusion, carbon burns only because the electro-negativity difference between carbon and oxygen produces an appreciable polarity which makes the bonds in the product more stable than the bonds in the reactants. The product is gaseous monomer because the lone pair bond weakening effect is greater in single bonds than in double bonds to oxygen, causing double bonds to be more stable than their equivalent in single bonds.

One more aspect of the combustion of carbon is the formation of CO not merely when there is insufficient oxygen but also at the temperatures of combustion, by the decomposition of carbon dioxide to CO and oxygen. First let us consider the bond energy in carbon monoxide. Since the bond is a triple bond, the lone pair bond weakening effect on oxygen must be completely eliminated, and the E''' energy of 104.0 kcal (435.1 kJ) is applicable. The geometric mean is therefore 94.2 kcal (394.1 kJ). The electronegativity of CO is 3.168, which corresponds to partial charges of 0.162 on C and -0.162 on O. The bond length is 112.8 pm. Using the triple bond factor of 1.787, we calculate the bond energy in CO:

$$E = 0.838 \times 94.2 \times 1.787 \times 147.4/112.8$$
$$+ 0.162 \times 33200 \times 1.787/112.8$$
$$= 184.3 + 85.2 = 269.5 \text{ kcal (1128 kJ) per mole}$$

The experimental value is 257.3 kcal (1079 kJ) per mole. The difference of about 12 kcal (51.0 kJ) may correspond to the energy of transferring one electron from oxygen to carbon, although this is only speculation.

Now consider the decomposition of CO_2. The experimental bond energy for each bond is 192.3 kcal (804.6 kJ) per mole, but that is not what it costs to break loose one oxygen atom. Here we must anticipate a major topic of this book to be discussed in Chapters 12 and 13. Dissociation of one oxygen liberates the electrons formerly used in bonding to it, allowing the second double bond to change to a triple bond which is 65.3 kcal stronger. The result is that the first bond is 65.3 kcal easier to break than anticipated from the contributing bond energy, requiring only about 127 kcal (531 kJ) per mole. Thus the decomposition of CO_2 at high temperatures occurs much more readily than we might have expected on the basis of the contributing bond energies.

WHY IS SILICON DIOXIDE SOLID?

A question unanswered until the advent of the quantitative theory of polar covalence is, why are carbon dioxide and silicon dioxide so different? Why carbon dioxide is monomeric and a gas has already been answered, so now the solid nature of SiO_2 must be explained. Let us first consider the nature of gaseous silica, formed when sand is vaporized at high temperature. The electronegativity is calculated to be 3.056, from which the partial charges are Si 0.400 and oxygen -0.199, giving an ionic blending coefficient of 0.300. The geometric mean homonuclear energy is 61.0 kcal (255.2 kJ) per mole. The bond length is 154 compared to a radius sum of 187.1 pm. From these data we can calculate the energy of an Si=O double bond:

$$E = 0.700 \times 61.0 \times 1.488 \times 187.1/154$$
$$+ 0.300 \times 33200 \times 1.488/154$$

= 77.2 + 96.2 = 173.4 kcal (725.5 kJ) per mole

Twice this or 346.9 kcal (1451 kJ) per mole is the calculated atomization energy. However, the experimental value is only 303.4 kcal (1269 kJ) per mole. As discussed earlier in this chapter, the size difference between oxygen and silicon atoms evidently prevents formation of normal double bonds between the two.

The strongest single bond that might form would involve the unweakened homonuclear bond energy of oxygen, 104.0 kcal (435.1 kJ) per mole, giving a geometric mean with silicon of 75.0 kcal (313.8 kJ). It could result from pairing of the oxygen electrons leaving a vacant orbital to accommodate a pair of silicon electrons: Si-O'''. The energy of such a bond would be:

E = 0.700 x 75.0 x 187.1/154 + 0.300 x 33200/154
= 63.8 + 64.7 = 128.5 kcal (537.6 kJ) per mole of bonds

Twice this is only 257 kcal, but adding 128.5 to the energy of one double bond, 173.4, gives 301.9 kcal (1213 kJ) per mole for the atomization energy, in good agreement with the experimental value of 303.4. The bonds in gaseous SiO_2 can be described as averaging Si=O" and Si-O'''. This kind of bond is not unique to this compound; other examples have been mentioned and will be given presently.

The next question is, how does the monomer compare with the polymer? Here again we must anticipate the following chapter, pointing out that there are many single bonds also in which the lone pair bond weakening is reduced, and the Si-O bond is one of these. The bond appears to be describable as Si-O", which removes any advantage a double bond might have over two single bonds. The total for this bond is calculated to be 446 kcal, close to the experimental value of the solid polymer of 444.8 kcal (1861 kJ) per mole. This is much higher even than the hypothetical O=Si=O energy of 346.9 kcal (1451 kJ) per mole. There is no question of the greater stability of the polymer.

The unique and valuable properties of the silicones have of course led to wondering why such carbon analogues as acetone do not undergo similar polymerization. The preceding examples provide an explanation: in carbon-oxygen compounds, double bonds are favored over their equivalent in single bonds wherever possible, whereas in silicon-oxygen compounds the reverse is true, because the lone pair bond weakening effect on the homonuclear energy of oxygen can be partly removed in single bonds by silicon but not by carbon, (see Chapter 5) eliminating any advantage of multiplicity in SiO bonds.

The instability of carbonic acid can similarly be explained. The decomposition changes two C-O bonds to one C=O bond with a gain in stability.

NITROGEN OXIDES

The conspicuous differences between oxides of nitrogen and oxides of the heavier M5 elements should lead every curious chemist to wonder why. Calculations similar to those detailed above (Tables 4:1, 4:3) show that multiple bonding in oxides of nitrogen is doubly favored over the equivalent in single bonds because multiple bonds decrease the lone pair bond weakening effects in the homonuclear energies of **both** nitrogen and oxygen. Furthermore the weakening in N is far greater than in P, As, and Sb, giving a greater advantage to N double bonds. This advantage in multiplicity is increased still more by the fact that the lone pair weakening effect affects only the covalent contribution to the total bond energy. Since nitrogen is far more electronegative than any of the other M5 elements, its bonds to oxygen are appreciably less polar, which means that the covalent contribution is relatively more important in the total bond energy. There is also the fact that nitrogen and oxygen atoms are nearly alike in size and easily form normal multiple bonds, whereas the heavier M5 atoms are too much larger than oxygen to allow this. All these factors taken together account very well for the large difference in oxygen chemistry between nitrogen and its heavier group members.

The bonding in nitrous oxide, N_2O, is especially curious. The atomization energy of 266.1 kcal (1113 kJ) per mole can be fairly well accounted for by assuming a normal double bond between the central nitrogen and the oxygen, N"=O". This leaves the central nitrogen atom incapable of forming a true triple bond to the terminal nitrogen atom, so the bond is apparently a single bond but with completely unweakened energy as in a triple bond: N'''-N'''. The calculated double bond energy is 144.8 kcal (605.8 kJ) and the single NN bond energy is 123.3 kcal (515.9 kJ) (since the bond length is as short as expected for a triple bond, 113 pm). The total is the sum, 268.1 kcal (1122 kJ), in reasonably good agreement with the 266.1 experimentally determined.

Again we must anticipate the emphasis of Chapter 12, and recognize that if the oxygen atom is split away from the nitrogen atoms, the latter can then form N_2 with a triple bond, having a dissociation energy of 225.9 kcal (945 kJ) per mole. In other words, the NN bond can gain strength by about 225.9 - 123.3 = 106.6 kcal (446 kJ) per mole of bonds. This means that breaking away the oxygen atom is 106.6 kcal easier than expected from the 144.8 kcal (619 kJ) contributing bond energy, and it costs only about 36 kcal (159 kJ) per mole to liberate the oxygen atom. An interesting property of N_2O is its ability to support combustion in the manner of O_2, and here is an explanation.

Why isn't the bond description more logically composed of a normal triple bond between the two nitrogen atoms and a single bond to the oxygen? With the weakest single bond possible--and note that the NO bond length is only 119 pm compared with the normal 144 pm--the calculated atomization energy is still about 20 kcal (84 kJ) per mole too high. In fact, the actual bonding is probably much more complex than indicated, but for purposes of calculation of the atomization energy, the first description is clearly superior.

SULFUR OXIDES

Since sulfur atoms and oxygen atoms are identical in outer structure, each having six outermost electrons, it might be expected that they could unite by a double bond, analogous to O_2 and S_2. The disparity in size evidently prevents this, for the bond is like that in gaseous SiO_2, a blend of a normal double bond and a coordination bond: S"=O" and S'-O'''. The calculated energy, the average of these two forms, is 122.2 kcal (511.3 kJ) per mole, compared to the experimental value of 124.7 kcal (521.7 kJ). The same kind of bond is observed in sulfur dioxide, except that they are a little stronger than in SO, being somewhat shorter, 143 pm compared to 149 in SO. The S"=O" energy is calculated to be 146.1 kcal (611.3 kJ) and the S'-O''' energy is 109.9 kcal (459.8 kJ), the sum being 256.0 kcal (1071 kJ) compared to the experimental value of 256.2 kcal (1071.9 kJ). It is therefore not surprising that SO is very unstable, rapidly disproportionating to SO_2 and elemental sulfur, for there is a gain in energy not merely for the SO bonds but also the equivalent of the atomization energy of the solid sulfur.

In SO_3 there is only one double bond that might be formed, with the other two being coordinate bonds. This evidently results in only a half reduction of the lone pair bond weakening effect, and the bonds are described as a blend of one double bond, S"=O", and two bonds, S'-O". These give a calculated energy of 340.7 kcal (1425 kJ) compared to the experimental value of 339.5 kcal (1420 kJ).

It is interesting that in S_2O, the double bond is between the two sulfur atoms, which are of course equal in size. The SO bond is like that in SO_2, an average of S"=O" and S'-O'''. Again disproportionation to sulfur and SO_2 is favored by the advantage of two single bonds over one double bond in solid sulfur, as well as by the increase in SO bonds.

Chapter 5

MORE ABOUT THE LONE PAIR BOND WEAKENING EFFECT ON SINGLE BONDS

In the preceding chapter were discussed many examples of reduction of the lone pair bond weakening effect (LPBWE) through the formation of multiple bonds. Examples of reduction of this effect in bonds where normal multiplicity is prevented by disparity in atomic size were also provided. It is now appropriate to examine this LPBWE in more detail, and especially as it relates to the single bonds formed by the elements of periodic groups M5, M6, and M7. Figure 5:1 shows the magnitude of the LPBWE in percentage of the unweakened bond energy.

THE ORIGIN OF LPBWE

Abnormally weak N-N, O-O, and F-F single bonds have long been recognized, and explained (A10, A11) as resulting from repulsions between lone pair electrons on the two bonded atoms. (See also A38 and A 39.) The development of the theory of polar covalence has not only permitted quantitative evaluation of this effect, but also provided convincing evidence that the **accepted explanation cannot be correct.** This evidence will be presented herein.

The LPBWE is so named because it is observed only in major group elements from M5 to M7, these being the only elements whose atoms have lone pair electrons in their valence shell. Although the exact mechanism by which the observed bond weakening occurs is not yet understood, the presence of **lone pair electrons** is the only obvious clue. It appears that such electrons, not being directly involved in the bonding, somehow interfere with the normal bonding forces, perhaps by reducing the effective nuclear charge sensed by the valence electrons. This supposition is enforced by the observation that in triple bonds, where the concentration of 6 electrons within the internuclear region could reasonably be expected to repel the lone pair electrons to the far side of the atoms opposite the bond, the weakening is removed completely. In double bonds, the effect is half removed. There are also, however, many **single** bonds in which the effect

69

Figure 5:1

Percentage LPBWE

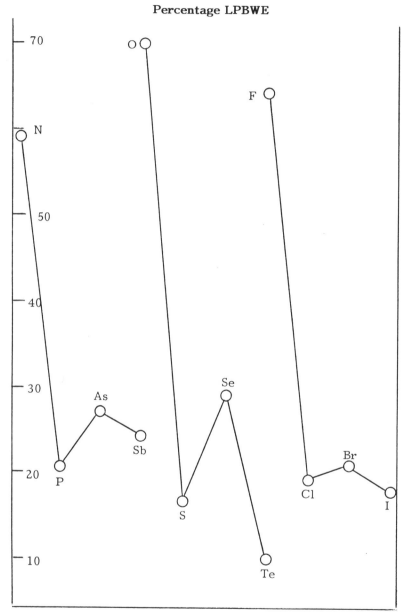

is either totally or half removed.

In Chapter 3 it was shown that the atomization energy of a compound can be determined as the sum of the contributing bond energies of all the bonds in the compound. In diatomic molecules or in molecules having bonds that are all alike, the contributing bond energy can be determined from the experimental standard heat of formation of the gaseous compound, by subtracting this from the sum of the atomization energies of all the atoms involved. If there are more than one kind of bond in the molecule, however, there is no experimental method of determining the individual contributing bond energies. Fortunately they can be calculated by the theory of polar covalence. If their sum equals the experimental atomization energy for the compound, it seems reasonable to assume that the individual contributing bond energies as calculated are probably correct. On this basis it is found that the fully weakened bond energies are applicable not merely to N-N, O-O, and F-F, but also to many other combinations of these elements. These include compounds with bonds to **hydrogen** and to **carbon,** neither of which has **any lone pair electrons.** How can the prevailing explanation, that repulsion between lone pairs on adjacent atoms weakens the bond between them, be correct when one of the bonded atoms has no such lone pair? The existence of LPBWE in these bonds can evidently be ascribed only to an internal effect, within the atom possessing the lone pair. **It is intra-atomic, not inter-atomic.**

The evidence is abundant. The nonpolar energy of a polar covalent bond is determined as the geometric mean of the two homonuclear bond energies, corrected for any difference between the sum of the nonpolar covalent radii and the experimental bond length. **The LPBWE affects only the nonpolar contribution** to the total bond, but this can be very significant. For example, the CBE (contributing bond energy) in H_2O, both experimental and calculated, is about 111 kcal (464 kJ) per mole of bonds. If the homonuclear bond energy of oxygen were unweakened, the CBE would be about 147 kcal (615 kJ) per mole of bonds, and the standard heat of formation of gaseous water would be -130 kcal (1544 kJ) instead of -58 kcal (-243 kJ) per mole as measured experimentally. This latter value is based on the **fully weakened homonuclear bond energy** of oxygen. This fully weakened oxygen energy is involved in all 22 ethers as well as in all hydroxyl compounds investigated. These latter include 25 alcohols, 16 phenols, 26 carboxylic acids, and 13 inorganic compounds including water and all common hydroxyacids—in all, 102 examples without exception. The fully weakened bond energy of fluorine is involved in HF and in many other inorganic fluorides such as NF_3, OF_2, and SF_6, including all solid fluorides which have been studied. The fully weakened bond energy of nitrogen is involved in ammonia and all amines, amides, and related compounds. Other examples are to be found in bonds of hydrogen and carbon to other M5, M6, and M7 elements. Nearly one hundred examples were presented in Chapter 3.

It must be concluded that the **LPBWE does not depend on repulsions between lone pair electrons on adjacent atoms, but results from lone pair interference with bonding by its own atom.**

OCCURRENCE OF LPBWE

For practically all molecules studied, the calculated atomization energy, using the fully weakened homonuclear single covalent bond energies, is either equal to or less than the experimental value. Wherever the calculated value was less than the experimental value, the possibility of reduced weakening in some of the bonds was investigated. This is not nearly as capricious a procedure as might be thought, for all generalizations are based on consistency in a number of similar compounds. The choice is usually quite clear, although occasionally somewhat arbitrary. For example, the average experimental bond energy in $AlF_3(g)$ is 141.5 kcal (592.0 kJ) per mole of bonds. The calculated values are, Al-F' 127.1 kcal (531.8 kJ), Al-F" 137.1 kcal (573.6 kJ), and Al-F''' 144.9 kcal (606.3 kJ) per mole of bonds. For gaseous $AlCl_3$, the experimental average bond energy is 101.8 kcal (425.9 kJ) per mole of bonds. The calculated values are, Al-Cl' 97.7 kcal (408.8 kJ), Al-Cl" 99.7 kcal (417.1 kJ), and Al-Cl''' 101.7 kcal (425.5 kJ) per mole of bonds. Assignment of Al-F''' and Al-Cl''' bond types seems appropriate, especially since similar bonds are observed in halides of B, Ga, and In. Table 5:1 summarizes the LPBWE in binary halides.

Table 5:1

LPBWE in Gaseous Halides

LiF'	BeF_2"		$\mathbf{BF_3}$'''	CF_4',"	NF_3'	OF_2'	F_2'
$LiCl'$	$BeCl_2'$		$\mathbf{BCl_3}$'''	CCl_4'	NCl_3'	Cl_2O'	ClF'
$LiBr'$	$BeBr_2'$		BBr_2"	CBr_4'	NBr_3'	Br_2O'	BrF'
LiI'	BeI_2		BI_3'	CI_4'	NI_3'		IF'
NaF'	MgF_2'		$\mathbf{AlF_3}$'''	$\mathbf{SiF_4}$'''	$\mathbf{PF_3}$'''	SF_4'	Cl_2'
$NaCl'$	$MgCl_2'$		$\mathbf{AlCl_3}$'''	$\mathbf{SiCl_4}$"	PCl_3'	SF_6'	$BrCl'$
$NaBr'$	$MgBr_2'$		$\mathbf{AlBr_3}$'''	$\mathbf{SiBr_4}$'''	PBr_3'	SCl_2'	ICl'
NaI'	MgI_2'		$\mathbf{AlI_3}$'''	SiI_4?	PI_3'		
KF'	$\mathbf{CaF_2}$'''	ZnF_2'	$\mathbf{GaF_3}$'''	$\mathbf{GeF_4}$'''	$\mathbf{AsF_3}$'''	SeF_6	Br_2'
KCl'	$\mathbf{CaCl_2}$'''	$ZnCl_2'$	$\mathbf{GaCl_3}$'''	$\mathbf{GeCl_4}$'''	$\mathbf{AsCl_3}$'''	$SeCl_2'$	IBr'
KBr'	$\mathbf{CaBr_2}$'''	$ZnBr_2'$	$\mathbf{GaBr_3}$'''	$\mathbf{GeBr_4}$'''	$\mathbf{AsBr_3}'$		
KI'	$\mathbf{CaI_2}$'''	ZnI_2'	$\mathbf{GaI_3}$'''	$\mathbf{GeI_4}$'''	$\mathbf{AsI_3}$'''		
RbF'	$\mathbf{SrF_2}$'''	CdF_2'	$\mathbf{InF_3}$'''	$\mathbf{SnF_4}$'''	$\mathbf{SbF_3}$'''	TeF_6'	I_2'
$RbCl'$	$\mathbf{SrCl_2}$'''	$CdCl_2'$	$\mathbf{InCl_3}$'''	$\mathbf{SnCl_4}$'''	$\mathbf{SbCl_3}$'''	$TeCl_4'$	
$RbBr'$	$\mathbf{SrBr_2}$'''	$CdBr_2'$	$\mathbf{InBr_3}$'''	$\mathbf{SnBr_4}$'''	$\mathbf{SbBr_3}$'''		
RbI'	$\mathbf{SrI_2}$'''	CdI_2'	$\mathbf{InI_3}$'''	$\mathbf{SnI_4}$'''	$\mathbf{SbI_3}$'''		
CsF'	$\mathbf{BaF_2}$'''	HgF_2'		PbF_4'	$\mathbf{BiF_3}$'''		
$CsCl'$	$\mathbf{BaCl_2}$'''	$HgCl_2'$		$PbCl_4'$	$\mathbf{BiCl_3}$'''		
$CsBr'$	$\mathbf{BaBr_2}$'''	$HgBr_2'$			$\mathbf{BiBr_3}$'''		
CsI'	$\mathbf{BaI_2}$'''	HgI_2			$\mathbf{BiI_3}$'''		

Because the LPBWE affects only the nonpolar covalent contribution to the bond energy, its importance diminishes with increasing bond polarity, making the bond type more difficult to recognize. The LPBWE's of nitrogen, oxygen, and fluorine are so large that there is relatively little difficulty in assignment of bond types involving these elements. For the heavier elements of these groups, assignment is not

Table 5:2

Bond Energies in Fluoride Molecules; LPBWE Diminished

cmpd	bond	R_o, pm	t_i	E_c	E_i	Ecalc	Eexp
BeF_2	Be-F"	140.	0.446	44.7	105.8	150.5	152.5
CaF_2	Ca-F'''	210.	0.729	18.4	115.3	133.7	134.
SrF_2	Sr-F'''	220.	0.881	10.5	125.4	135.9	129.8
BaF_2	Ba-F'''	232.	0.867	7.6	124.1	131.7	137.9
BF_3	B-F'''	131.	0.338	71.5	86.7	158.2	154.0
AlF_3	Al-F'''	163.	0.493	44.5	100.4	144.9	141.5
GaF_3	Ga-F'''	188.	0.302	48.5	53.3	101.8	
InF_3	In-F'''	216.	0.370	38.1	56.9	95.0	
CF_4	C-F'	132.	0.232	49.7	58.4	108.1	
	C-F"	132.	0.232	68.5	58.4	126.9	
					aver	117.5	117.0
SiF_4	Si-F'''	154.	0.375	58.7	80.8	139.5	142.4
GeF_4	Ge-F'''	168.	0.259	59.8	51.2	110.0	112.5
SnF_4	Sn-F'''	186.	0.334	47.8	59.6	107.4	
PF_3	P-F'''	157.	0.279	64.0	59.0	123.0	120.3
PF_5	P-F'	157.	0.286	37.9	60.5	98.4	
	P-F'''(3)	157.	0.286	63.3	60.5	123.8	
						113.6	111.1
AsF_3	As-F'''	171.2	0.215	58.5	41.7	100.2	105.6
SbF_3	Sb-F'''	190.	0.293	50.8	51.2	102.0	103.0
BiF_3	Bi-F'''	196.	0.319	45.1	54.0	99.1	
BHF_2	B-H	115.	0.074	73.2	21.4	94.6	
	B-F'''	130.	0.319	73.3	81.5	154.8	
						404.2	399.1
BBr_2F	B-Br"	187.	0.196	53.4	34.8	88.2	
	B-F'''	130.	0.319	73.3	81.5	154.8	
						331.3	329.2
B_2F_4	B-B	167.	0.0	75.5	0.0	75.5	
	B-F'''	132.	0.329	71.1	82.7	153.8	
						690.9	685.4
BF_2Cl	B-F'''	130.	0.331	72.0	84.5	156.5	
	B-Cl"	175.	0.249	53.5	47.2	100.7	
						413.7	413.4
$BFCl_2$	B-F'''	130.	0.325	72.7	83.0	155.7	
	B-Cl'''	175.	0.244	58.2	46.3	104.5	
						364.7	364.9
BF_2OH	H-O'	96.	0.197	42.3	68.1	110.4	
	B-O"	136.	0.272	59.2	66.4	125.6	
	B-F'''	130.	0.324	72.8	82.7	155.5	
						547.1	544.2
CF_3Cl	C-F'(2)	132.8	0.228	49.7	57.0	106.7	
	C-F"	132.8	0.228	68.4	57.0	125.4	
	C-Cl'	175.1	0.143	60.8	27.1	87.9	
						426.7	428.3

Table 5:2 cont.

cmpd	bond	R_o, pm	t_i	E_c	E_i	Ecalc	Eexp
CF_2Cl_2	C-F'	133.8	0.225	49.5	55.8	105.3	
	C-Cl'	177.5	0.141	60.2	26.4	86.6	
						383.7	386.2
CF_3Br	C-F'(2)	133.0	0.227	49.7	56.7	106.4	
	C-F"	133.0	0.223	68.4	56.7	125.1	
	C-Br'	190.8	0.096	56.9	16.7	73.6	
						411.4	412.7
CF_3I	C-F'(2)	133.2	0.223	49.8	55.6	105.4	
	C-F"	133.2	0.223	68.7	55.6	124.3	
	C-I'	213.4	0.007	54.4	1.1	55.5	
						395.2	397.5
SiH_3F	Si-H	147.	0.102	66.5	23.0	89.5	
	Si-F"	159.	0.331	50.2	69.1	119.3	
						387.8	388.2
SiH_2F_2	Si-H	147.	0.107	62.7	24.2	86.9	
	Si-F'''	158.	0.345	60.0	72.5	132.5	
						438.7	444.0
$SiHF_3$	Si-H	146.	0.112	58.7	25.5	84.2	
	Si-F'''	156.	0.361	59.3	76.8	136.1	
						492.6	499.8

as easy, and justifies confidence only when consistency over a number of similar combinations is observable. With these limitations in mind, a number of generalizations can be made, based on data previously presented in Chapters 3 and 4 and in Tables 5:2-5:11.

1) In single bonds, the reduction of weakening occurs only in the atom bearing partial negative charge, or in the more negative of the bonding pair, even where both atoms are possibly subject to LPBWE. (In multiple bonds, where both atoms are susceptible to bond weakening, both exhibit reduction of the weakening.)

2) LPBWE reduction is never observed in nonmolecular solids, only, if at all, in the gaseous molecules.

3) Reduction of LPBWE rarely occurs except where the other atom has at least one vacant orbital available. However, this may be an outer d orbital, not necessarily being

Table 5:3

Bond Energies in Chloride Molecules; LPBWE Diminished

cmpd	bond	R_o, pm	t_i	E_c	E_i	Ecalc	Eexp
$CaCl_2$	Ca-Cl'''	251.	0.627	19.1	83.0	102.1	106.7
$SrCl_2$	Sr-Cl'''	267.	0.726	12.5	90.3	102.8	106.6
$BaCl_2$	Ba-Cl'''	282.	0.762	10.0	89.7	99.7	113.1
BCl_3	B-Cl'''	175.	0.240	58.5	45.8	104.3	105.7
$AlCl_3$	Al-Cl'''	206.	0.386	39.5	62.2	101.7	101.8
$GaCl_3$	Ga-Cl'''	209.	0.206	46.0	32.7	78.7	
$InCl_3$	In-Cl'''	246.	0.272	35.4	36.7	72.1	73.8
$SiCl_4$	Si-Cl'''	201.9	0.277	48.3	45.5	93.8	95.3
$GeCl_4$	Ge-Cl'''	209.	0.166	50.2	26.4	76.6	78.2
$SnCl_4$	Sn-Cl'''	231.	0.220	41.4	31.6	73.0	75.2
$AsCl_3$	As-Cl'''	216.1	0.122	48.2	18.7	66.9	
$SbCl_3$	Sb-Cl'''	232.5	0.197	43.3	28.1	71.4	75.1
$BiCl_3$	Bi-Cl'''	248.	0.224	37.1	30.0	67.1	66.7
$BHCl_2$	B-H	113.	0.071	78.6	20.9	99.5	
	B-Cl'''	175.	0.231	59.2	43.8	103.0	
						305.5	303.2
BBr_2Cl	B-Br''	187.	0.193	53.8	34.3	88.1	
	B-Cl'''	175.	0.235	58.9	44.6	103.5	
						279.7	281.3
B_2Cl_4	B-B	175.	0.0	72.1	0.0	72.1	
	B-Cl'''	173.	0.235	59.6	45.1	104.7	
						490.9	500.8
$SiHCl_3$	Si-H	148.	0.107	62.2	24.0	86.2	
	Si-Cl''	202.	0.268	46.5	44.0	90.5	
						357.8	359.1
PCl_5	P-Cl'(½)	219.	0.187	43.5	28.3	71.8(2)	
	P-Cl''(3)	204.	0.187	49.4	30.4	79.8	
						311.3	314.

confined to the octet. For example, in solid SiO_2 the bonds may be represented as Si-O".

4) Carbon provides the only recognized exception to this rule. In CF_4, the bonds appear to be an average of C-F' and C-F". In all carboxylic acids and esters, the single bond from carbonyl carbon to bridging oxygen involves the O" energy. In acyl halides except the iodides, the halogen energies are, F", F''' average, Cl", and Br". In amides, the CN bond appears to be an average of C-N' and C-N".

Table 5:4

Bond Energies in Bromide Molecules; LPBWE Diminished

cmpd	bond	R_o, pm	t_i	E_c	E_i	Ecalc	Eexp
$CaBr_2$	Ca-Br'''	263.	0.574	19.7	72.5	92.2	95.
$SrBr_2$	Sr-Br'''	282.	0.673	13.4	79.2	92.6	95.3
$BaBr_2$	Ba-Br'''	299.	0.708	11.0	78.6	88.6	98.8
BBr_3	B-Br''	187.	0.191	53.8	33.9	87.7	87.6
$AlBr_3$	Al-Br'''	227.	0.333	37.4	48.7	86.1	85.7
$GaBr_3$	Ga-Br'''	228.	0.158	43.0	23.0	66.0	
$InBr_3$	In-Br'''	258.	0.222	34.4	28.6	63.0	64.2
$SiBr_4$	Si-Br'''	216.	0.226	46.5	34.7	81.2	78.8
$GeBr_4$	Ge-Br'''	230.	0.118	46.4	17.1	63.5	67.1
$SnBr_4$	Sn-Br'''	244.	0.187	38.5	25.2	63.7	63.6
$AsBr_3$	As-Br'''	233.	0.076	45.2	10.8	56.0	61.2
$SbBr_3$	Sb-Br'''	251.	0.150	40.6	19.8	60.4	63.3
$BiBr_3$	Bi-Br'''	263.	0.175	35.5	22.1	57.6	56.0
$BHBr_2$	B-H	120.	0.070	75.8	19.4	95.2	
	B-Br''	187.	0.185	54.2	32.8	87.0	
						269.2	264.4

5) Not one of the 21 boron compounds studied (Tables 5:7, 5:8, and 5:10) which contain single bonds between boron and oxygen exhibits the completely weakened oxygen homonuclear energy, the O'' energy being involved in most.

6) In 14 compounds containing P-O single bonds (Table 5:9), not one exhibits the fully weakened oxygen energy, all being either P-O'' or P-O'''.

7) Of 36 compounds containing S-O single bonds (Tables 5:7, 9:3-9:5), either terminal or bridging, not one involves the fully weakened oxygen energy, all being either S-O'' or S-O'''.

8) Fully weakened O' energy is involved in all hydroxyl groups and in OF_2, Cl_2O, HNO_2, HNO_3, NH_2OH, $HClO_4$, and in all carbon-oxygen bonds in alcohols, ethers, and other alkoxy compounds.

9) Binary halides having unweakened bonds are listed in Tables 5:1-5:4. Here occur most of the larger discrepancies between calculated and experimental bond energies, for,

Table 5:5

Bond Energies in Iodide Molecules; LPBWE Diminished

cmpd	bond	R_o, pm	t_i	E_c	E_i	Ecalc	Eexp
CaI_2	Ca-I'''	288.	0.480	20.4	55.5	75.9	78.3
SrI_2	Sr-I'''	303.	0.576	13.5	63.1	76.6	77.6
BaI_2	Ba-I'''	320.	0.611	12.2	66.9	79.1	85.8
AlI_3	Al-I'''	244.	0.243	37.0	33.1	70.0	68.1
GaI_3	Ga-I'''	245.8	0.072	41.2	9.7	50.9	
InI_3	In-I'''	280.	0.134	32.9	15.9	48.8	50.2
GeI_4	Ge-I'''	249.	0.032	44.1	4.3	48.4	51.4
SnI_4	Sn-I'''	264.	0.099	37.3	12.5	49.8	
AsI_3	As-I'''	253.	0.007	42.1	0.9	43.0	
SbI_3	Sb-I'''	271.9	0.064	38.5	7.8	46.3	
BiI_3	Bi-I'''	284.	0.089	33.8	10.4	44.2	43.3

on the average for the thousand-plus compounds studied, the agreement is within 1 kcal (4 kJ) per mole of bonds. Taken together with the high polarity of some of these bonds, the discrepancies make the assignment of triple prime unweakened homonuclear energies here somewhat speculative.

10) Halogen compounds in which the fully weakened homonuclear bond energy of halogen is involved include all organic monochlorides, monobromides, and monoiodides, gaseous halides of the M1 elements and Be, Mg, N, O, S, Se, and Te, and interhalogen compounds. They also include PCl_3, PBr_3, and PI_3, gaseous halides of M2' (Zn, Cd, and Hg), all the hydrogen halides, BI_3, the thionyl halides, $POBr_3$, AlOCl, $COCl_2$, cyanogen halides, hypohalous acids, NO_2F and NO_2Cl, and gaseous halides of Cu(I) and Ag(I).

11) In 29 compounds containing bonds of boron to fluorine, chlorine, or bromine, all have B-X''' or B-X'' bonds except, apparently, BOCl' and BOBr'.

12) Table 5:11 lists a number of hydrohalides and oxyhalides, which illustrate a tendency, where similar compounds are compared, for the removal of weakening to be enhanced as the positive charge on the "normalizing" atom (the atom that reduces the weakening) increases.

Table 5:6

Bond Energies in O–containing Molecules; LPBWE Diminished

cmpd	bond	R_o, pm	t_i	E_c	E_i	Ecalc	Eexp
$B(OH)_3$	H-O'	96.	0.188	45.6	65.0	110.6	
	B-O"	136.	0.260	59.7	63.5	123.2	
						701.4	706.4
P_4O_6	P'-O"	165.	0.212	52.5	42.7	95.2	
	P'-O'''	165.	0.212	64.5	42.7	107.2	
					aver	101.2	100.4
P_4O_{10}	P'-O"(4)	139.	0.216	62.0	51.6	113.6	
	P'-O"(12)	162.	0.216	53.2	44.3	97.5	
						1624.4	1605.0
As_4O_6	As-O"	179.	0.151	47.7	28.0	75.7	77.9
H_2SO_4	H-O'	96.	0.197	43.9	68.1	112.0	
	S-O"	153.	0.121	61.9	26.3	88.2	
	S-O"	142.	0.121	66.7	28.3	95.0	
						590.3	585.6
Cl_2O_7	Cl-O"(@)	172.	0.031	60.4	6.0	66.4	
	Cl-O'	142.	0.031	51.0	7.3	58.3	
	Cl-O"	142.	0.031	73.1	7.3	80.4	
						549.2	543.5

13) Single bonds of boron to sulfur are B-S', as in boron thioalkyls, and to nitrogen are B-N", as in $B_3N_3H_6$ and its derivatives.

14) Where disparity in atomic radius evidently prevents normal double bonding, the bond can commonly be described, in terms of energy, as an average of a normal double bond and a coordinated unweakened single bond: S"=O", S'-O''', as in SO and SO_2. Similar bonds are recognized in gaseous AlOF, AlOCl, SiO_2, and, surprisingly, in COFCl and $COCl_2$.

A consistent interpretation of all these observations, in terms of atomic fundamentals, has not yet been developed. It is interesting that the availability of just one vacant orbital, as in boron, for example, can provide the means of removing the LPBWE in three halogen atoms simultaneously. Furthermore, the phenomenon does not involve the actual sharing of additional electrons, for this would be reflected in a multiplicity factor applicable to the entire bond energy, not

Table 5:7

Bond Energies in Oxyhalide and Thiohalide Molecules

cmpd	bond	R_o, pm	t_i	E_c	E_i	Ecalc	Eexp
POF_3	P'-O"	145.	0.227	58.6	52.0	110.6	
	P'-F'''	152.	0.281	65.9	61.4	127.3	
						492.5	485.2
POF_2Cl	P'-O"	155.	0.223	55.2	47.8	103.0	
	P'-F'''	151.	0.277	66.7	60.9	127.6	
	P'-Cl"	201.	0.193	47.1	31.9	79.0	
						437.2	433.8
$POCl_3$	P'-O"	145.	0.216	59.5	49.5	109.0	
	P'-Cl'''	199.	0.187	53.3	31.2	84.5	
						362.5	359.5
$POBr_3$	P'-O"	145.	0.211	59.8	48.3	108.1	
	P'-Br'	206.	0.140	48.0	22.6	70.6	
						319.9	316.3
$PSCl_3$	P'-Cl'''	202.	0.184	52.7	30.2	82.9	
	P'-S'''	185.	0.092	63.0	16.5	79.5	
						328.2	327.7
SOF_2	S'-O'''	141.	0.129	81.8	30.4	112.2	
	S'-F'	153.	0.184	43.5	39.9	83.4	
						279.0	276.6
SO_2F_2	S'-O'''	141.	0.129	81.8	30.4	112.2	
	S'-F"	153.	0.184	59.9	39.9	99.8	
						423.9	428.1
$SOCl_2$	S'-O'''	145.	0.124	80.0	28.4	108.4	
	S'-Cl'	207.	0.095	50.4	15.2	65.6	
						239.6	234.6
SO_2Cl_2	S'-O"	143.	0.125	65.8	29.0	94.8	
	S'-Cl"	203.	0.095	54.4	15.5	69.9	
						329.4	330.3
$S_2O_5Cl_2$	S'-O"	145.	0.125	65.0	28.6	93.6	
	S'-O'(2)	153.	0.125	43.0	27.1	70.1	
	S'-Cl"	207.	0.096	48.5	15.4	63.9	
						642.4	641.4
$SOBr_2$	S'-O"	145.	0.122	65.1	27.9	93.0	
	S'-Br'	218.	0.049	48.1	7.5	55.6	
						204.2	201.1

Table 5:8

Bond Energies in Single-Bonded Rings; LPBWE Diminished

cmpd	bond	R_o, pm	t_i	E_c	E_i	Ecalc	Eexp
$(BOF)_3$	B-O"	136.	0.272	59.2	66.4	125.6	
	B-F'''	130.	0.324	72.8	82.7	155.5	
						1220.3	1202.2
$B_3O_3FH_2$	B-H	114.	0.071	77.6	20.7	98.3	
	B-O"	136.	0.259	60.3	63.2	123.5	
	B-F'''	130.	0.308	74.5	78.7	153.2	
						1091.0	1086.2
$B_3O_3F_2H$	B-H	114.	0.073	74.8	21.3	96.1	
	B-O"	136.	0.265	59.8	64.7	124.5	
	B-F'''	130.	0.316	73.6	80.7	154.3	
						1151.7	1146.0
$(BOCl)_3$	B-O"	136.	0.265	59.8	64.7	124.5	
	B-Cl"	175.	0.237	55.9	45.0	100.9	
						1049.7	1057.2
$(BOCl)_4$	B-O"	136.	0.265	59.8	64.7	124.5	
	B-Cl"	175.	0.237	55.9	45.0	100.9	
						1399.6	1391.6
$B_3O_3ClH_2$	B-H	114.	0.071	78.4	20.7	99.1	
	B-O"	136.	0.257	60.4	62.8	123.2	
	B-Cl"	175.	0.229	56.5	43.4	99.9	
						1038.4	1027.3
$B_3O_3Cl_2H$	B-H	114.	0.072	76.6	21.0	97.6	
	B-O"	136.	0.261	60.1	63.7	123.8	
	B-Cl"	175.	0.233	56.2	44.2	100.4	
						1041.2	1029.2
$(HOBO)_3$	H-O'	96.	0.189	45.3	65.4	110.7	
	B-O"	136.	0.261	60.1	63.7	123.8	
						1446.3	1459.0
$(BNH_2)_3$	B-H	120.	0.068	78.5	18.8	97.3	
	N-H	102.	0.108	58.0	35.2	93.2	
	B-N"	144.	0.176	63.8	40.6	104.4	
						1197.7	1177.4
$(BNClH)_3$	B-N"	144.	o.183	63.2	42.2	105.4	
	N-H	102.	0.108	55.3	36.8	92.1	
	B-Cl'''	175.	0.228	59.4	43.3	102.7	
						1216.6	1221.5
$(PNCl_2)_3$	P'-Cl"	197.	0.183	51.4	30.8	82.2	
	P'-N'	165.	0.135	44.0	27.2	71.2	
						920.5	915.1

Table 5:9

Bond Energies in Alkoxyboron Molecules; LPBWE Diminished

cmpd	bond	R_o, pm	t_i	E_c	E_i	Ecalc	Eexp
$(CH_3O)_3B$	B-O"	136.	0.252	60.9	61.5	122.4	
	B-O'''	136.	0.252	74.9	61.5	136.4	
	C-O'	143.	0.152	46.1	35.3	81.4	
	(CH_3)					296.4	
						1521.6	1509.9
$(CH_3O)_2BH$	B-H	114.	0.069	75.9	20.1	96.0	
	B-O"	136.	0.251	60.9	61.3	122.2	
	B-O'''	136.	0.251	74.9	61.3	136.2	
	C-O'	143.	0.151	46.1	35.1	81.2	
	(CH_3)					296.4	
						1109.6	1098.6
$(CH_3O)_2BCl$	B-Cl'''	175.	0.227	59.5	43.1	102.6	
	B-O"	136.	0.254	60.7	62.0	122.7	
	B-O'''	136.	0.254	74.7	62.0	136.7	
	C-O'	143.	0.153	46.1	35.5	81.6	
	(CH_3)					296.4	
						1118.0	1115.1
$(C_2H_5O)_3B$	B-O"	136.	0.250	61.0	61.0	122.0	
	B-O'''	136.	0.250	75.1	61.0	136.1	
	C-O'	143.	0.159	45.7	36.9	82.6	
	(C_2H_5)					576.8	
						2365.4	2361.7
$(C_2H_5O)_2BCl$	B-Cl'''	175.	0.224	59.8	42.5	102.3	
	B-O'''	136.	0.251	75.0	61.3	136.3	
	C-O'	143.	0.152	46.1	35.3	81.4	
	(C_2H_5)					576.8	
						1691.2	1684.0
$(C_2H_5O)BCl_2$	B-Cl'''	175.	0.227	59.5	43.1	102.6	
	B-O'''	136.	0.254	74.7	62.0	136.7	
	C-O'	143.	0.153	46.1	35.5	81.6	
	(C_2H_5)					576.8	
						1000.3	1003.3
$(C_3H_7O)_3B$	B-O"	136.	0.249	61.1	60.8	121.9	
	B-O'''	136.	0.249	75.2	60.8	136.0	
	C-O'	143.	0.150	46.2	34.8	81.0	
	(C_3H_7)					857.2	
						3201.5	3205.9
$(C_4H_9O)_3B$	B-O"	136.	0.249	61.1	60.8	121.9	
	B-O'''	136.	0.249	75.2	60.8	136.0	
	C-O'	143.	0.150	46.2	34.8	81.0	
	(C_4H_9)					1137.6	
						4042.8	4049.0

Table 5:10

Bond Energies in Alkyl and Alkoxy Phosphorus Molecules LPBWE Diminished

cmpd	bond	R_o, pm	t_i	E_c	E_i	Ecalc	Eexp
CH_3POCl_2	C-P	184.	0.048	66.0	8.7	74.7	
	P-O'''	145.	0.204	74.2	46.7	120.9	
	P-Cl'''	199.	0.176	54.0	29.4	83.4	
	(CH_3)					296.4	
						658.8	657.7
$C_2H_5POCl_2$	C-P	184.	0.048	66.0	8.7	74.7	
	P-O'''	145.	0.201	74.5	46.0	120.5	
	P-Cl'''	199.	0.174	54.1	29.0	83.1	
	(C_2H_5)					576.8	
						938.2	936.5
$P(OCH_3)_3$	P-O'	157.	0.199	39.2	42.1	81.3	
	P-O''	157.	0.199	56.1	42.1	89.8	
	C-O'	143.	0.152	46.1	35.3	81.4	
	(CH_3)					296.4	
						1402.8	1409.2
$P(OC_2H_5)_3$	P-O''	157.	0.197	56.3	41.7	98.0	
	C-O'	143.	0.151	46.1	35.3	81.4	
	(C_2H_5)					576.8	
						2263.6	2263.2
$P(iC_3H_7)_3$	P-O''	157.	0.197	56.3	41.7	98.0	
	C-O'	143.	0.151	46.1	35.3	81.4	
	(iC_3H_7)					857.2	
						3114.3	3117.2
$OP(OC_2H_5)_3$	P-O''	147.	0.199	59.9	44.9	104.8	
	P-O'''(3)	155.	0.199	69.8	42.6	112.4	
	C-O'	143.	0.152	46.1	35.3	81.4	
	(C_2H_5)					576.8	
						2416.6	2411.5
$OP(OC_3H_7)_3$	P-O''	147.	0.197	60.1	44.5	104.6	
	P-O'''(3)	155.	0.197	69.8	42.6	112.4	
	C-O'	143.	0.152	46.1	35.3	81.4	
	(C_3H_7)					857.2	
						3257.6	3252.3
$OP(OC_4H_9)_3$	P-O''	147.	0.197	60.1	44.5	104.6	
	P-O'''(3)	155.	0.197	69.8	42.6	112.4	
	C-O'	143.	0.151	46.1	35.3	81.4	
	(C_4H_9)					1137.6	
						4098.8	4110.4

Table 5:11

Bond Energies in Alkyl and Thioalkyl Boron Molecules
LPBWE Diminished

cmpd	bond	R_o, pm	t_i	E_c	E_i	Ecalc	Eexp
$(C_4H_9)_2BOH$							
	H-O'	96.	0.179	51.1	61.9	113.0	
	B-O"	136.	0.247	61.3	60.3	121.3	
	B-C	156.	0.098	73.4	20.9	94.3	
	(C_4H_9)					1137.6	
						2698.1	2685.0
$(C_4H_9)_2BCl$							
	B-Cl'''	174.	0.220	60.4	42.0	102.4	
	B-C	156.	0.098	73.4	20.9	94.3	
	(C_4H_9)					1137.6	
						2566.2	2558.4
$(C_4H_9)_2BBr$							
	B-Br"	187.	0.180	54.5	32.0	86.5	
	B-C	156.	0.098	73.4	20.9	94.3	
	(C_4H_9)					1137.6	
						2550.3	2540.5
$(CH_3S)_3B$							
	B-S'	176	0.137	59.5	25.6	85.1	
	C-S'	182.	0.039	65.9	7.1	73.0	
	(CH_3)					296.4	
						1363.6	1352.1
$(C_2H_5S)_3B$							
	B-S'	176.	0.136	59.5	25.7	85.2	
	C-S'	182.	0.038	65.9	6.9	72.8	
	(C_2H_5)					576.8	
						2204.4	2208.4
$(C_3H_7S)_3B$							
	B-S'	176.	0.136	59.5	25.7	85.2	
	C-S'	182.	0.038	65.9	6.9	72.8	
	(C_3H_7)					857.2	
						3045.6	3047.7
$(C_4H_9S)_3B$							
	B-S'	176.	0.136	59.5	25.7	85.2	
	C-S'	182.	0.038	65.9	6.9	72.8	
	(C_4H_9)					1137.6	
						3886.8	3880.4

Table 5:12

Bonds between Elements and Examples of Each in Gaseous Inorganic Molecules Studied

H-F 1	Cu-I 1	Hg-F 1	Si-F 4	P-O 17
H-Cl 1	Ag-F 1	Hg-Cl 1	Si-Cl 4	P-S 1
H-Br 1	Ag-Cl 1	Hg-Br 1	Si-Br 2	As-F 1
H-I 1	Ag-Br 1	Hg-I 1	Si-I 1	As-Cl 1
H-O 14	Ag-I 1	B-F 10	Si-O 2	As-Br 1
H-S 2	Be-F 1	B-Cl 15	Si-S 1	As-I 1
H-Se 1	Be-Cl 1	B-Br 7	Si-Se 1	As-O 2
H-Te 1	Be-Br 1	B-I 2	Si-N 1	As-As 1
H-N 3	Be-I 1	B-O 22	Si-Si 1	Sb-F 1
H-P 2	Be-O 1	B-S 6	Ge-F 1	Sb-Cl 1
H-As 1	Be-S 1	B-N 2	Ge-Cl 1	Sb-Br 1
H-Sb 1	Mg-F 1	B-C 4	Ge-Br 1	Sb-I 1
H-C 22	Mg-Cl 1	B-B 2	Ge-I 1	Sb-O 1
H-Si 7	Mg-Br 1	Al-F 2	Ge-O 1	Sb-Sb 1
H-Ge 3	Mg-I 1	Al-Cl 2	Ge-S 1	Bi-F 1
H-B 9	Mg-O 1	Al-Br 1	Ge-Se 1	Bi-Cl 1
Li-F 1	Ca-F 1	Al-I 1	Ge-Te 1	Bi-Br 1
Li-Cl 1	Ca-Cl 1	Al-O 3	Ge-Ge 3	Bi-I 1
Li-Br 1	Ca-Br 1	Al-S 1	Sn-F 1	Bi-O 1
Li-I 1	Ca-I 1	Ga-F 1	Sn-Cl 1	Bi-Bi 1
Na-F 1	Ca-O 1	Ga-Cl 1	Sn-Br 1	O-F 2
Na-Cl 1	Sr-F 1	Ga-Br 1	Sn-I 1	O-Cl 7
Na-Br 1	Sr-Cl 1	Ga-I 1	Sn-O 1	O-O 2
Na-I 1	Sr-Br 1	Ga-O 1	Sn-S 1	O-S 13
K-F 1	Sr-I 1	In-F 1	Pb-F 1	O-Se 1
K-Cl 1	Sr-O 1	In-Cl 1	Pb-Cl 1	O-Te 1
K-Br 1	Ba-F 1	In-Br 1	Pb-Br 1	S-F 4
K-I 1	Ba-Cl 1	In-I 1	Pb-I 1	S-Cl 4
Rb-F 1	Ba-Br 1	In-O 1	Pb-O 1	S-Br 1
Rb-Cl 1	Ba-I 1	C-F 8	Pb-S 1	S-S 1
Rb-Br 1	Ba-O 1	C-Cl 5	N-F 2	S-Se 1
Rb-I 1	Zn-F 1	C-Br 2	N-O 8	Se-F 1
Cs-F 1	Zn-Cl 1	C-I 2	N-S 1	Se-Cl 1
Cs-Cl 1	Zn-Br 1	C-O 20	N-N 2	Se-Se 1
Cs-Br 1	Zn-I 1	C-S 7	P-F 6	Te-Te 1
Cs-I 1	Cd-F 1	C-Se 1	P-Cl 9	F-Cl 2
Cu-F 1	Cd-Cl 1	C-N 5	P-Br 2	F-Br 1
Cu-Cl 1	Cd-Br 1	C-P 2	P-I 1	F-I 1
Cu-Br 1	Cd-I 1	C-C 17	P-P 1	Cl-Br 1
				Cl-I 1

Table 5:13

Summary of Gaseous Inorganic Bonds Studied

(A) No. of examples of bonds to (B) no. of different elements.

	A	B		A	B		A	B		A	B
H	70	16	Ca	5	5	C	91	11	Bi	6	6
Li	4	4	Sr	5	5	Si	17	9	O	126	22
Na	4	4	Ba	5	5	Ge	11	9	S	39	13
K	4	4	Zn	4	4	Sn	6	6	Se	6	6
Rb	4	4	Cd	4	4	Pb	6	6	Te	3	3
Cs	4	4	Hg	4	4	N	20	6	F	68	37
Cu	4	4	B	70	9	P	40	8	Cl	73	35
Ag	4	4	Al	10	6	As	8	7	Br	40	32
Be	6	6	Ga	5	5	Sb	7	7	I	33	29
Mg	5	5	In	5	5						

just the covalent contribution. These facts suggest that the function of the "normalizing" atom which diminishes the weakening is somehow to distract the lone pair electrons on the other atom from their bond weakening position. Nearly always this involves a partial positive charge on the normalizing atom. However, in 20 alkyl derivatives: 2 sulfoxides, 15 sulfones, 5 sulfites, and 4 sulfates, in this study, the partial charge on sulfur that would result from electronegativity equalization is slightly negative, yet the bonds to oxygen are S-O" bonds. A simple explanation is so far elusive.

Despite the present lack of satisfactory explanation, the LPBWE phenomenon is established empirically by abundant experimental data and it has a very significant influence on the strength and nature of many chemical bonds.

SUMMARY

The bond study summarized within the three chapters, 3, 4, and 5, includes about 300 inorganic molecules containing

Table 5:14

**Gaseous Inorganic Molecules Showing Poor Agreement between
Calculated and Experimental Bond or Atomization Energy**

	BE or At E				BE or At E		
cmpd	calc	exp	% diff	cmpd	calc	exp	% diff
HI	67.1	71.4	6.0	ICl	58.1	50.5	15.1
LiF	170.4	138.±1	22.6	$HClO_4$	336.0	318.0	5.7
NaF	143.1	114.5	24.4	$SeCl_2$	56.7	60.0	5.5
KF	134.4	119.	12.9	NaBr	98.2	88.5±3	7.3
RbF	136.8	124.±3	8.0	$BaBr_2$	88.6	98.8	10.3
CsF	138.9	127.±4	6.0	$GeBr_4$	63.5	67.1	5.4
CuF	79.8	91.5	12.8	$PbBr_4$	51.8	49.4	4.9
SrF_2	135.9	129.8	4.7	$AsBr_3$	56.0	61.2	8.5
BaF_2	131.7	137.9	4.5	$SbBr_3$	60.4	63.3	4.6
OF_2	47.2	44.9	5.1	IBr	48.3	42.8	12.9
SF_4	82.4	78.6	4.8	NaI	82.4	72.±2	11.4
SF_6	82.5	78.6	5.0	CuI	54.3	47.±5	4.4
LiCl	124.1	112.±3	7.9	BaI_2	66.9	79.1	7.8
NaCl	109.6	98.±2	9.6	HgI_2	32.3	34.6	6.7
CuCl	79.8	91.5	12.8	GeI_4	48.4	51.4	5.8
$CaCl_2$	102.1	106.7	5.2	BeO	126.5	107.±5	13.0
$BaCl_2$	99.7	113.1	11.9	CaO	123.8	119.5±3.5	8.8
$SbCl_3$	71.4	75.1	4.9	SrO	129.7	109.±4	15.5
SCl_2	67.7	64.8	4.5	CO	269.5	257.3	4.7
BrCl	56.3	52.3	7.7	GeS	121.9	131.5	7.0
				GeTe	84.9	96.±2	9.7

1384 bonds. These contain all the major group elements except
thallium, in essentially all their compounds for which heats
of formation for the gaseous state could be found. They
also include (I) halides of copper and silver, which can reasonably be included in the major group chemistry in the (I) state,
and of course, zinc, cadmium and mercury are major group
elements as M2' since their chemistry never involves the
underlying d orbitals. Table 5:12 lists the various combinations
of elements studied and the number of different molecules
containing each kind of bond. No subdivision according to
multiplicity or LPBWE is provided here. There are listed,
in Table 5:12, 198 different combinations of elements. Table
5:13 further summarizes these data by listing for each element
the number of other elements with which its bonds have been

studied, and the total number of examples of molecules containing these bonds.

Literature values for the heats of formation of 290
of these gaseous molecules are available to allow comparison
of calculated with experimental atomization energies for
these compounds. The average difference for the 290 compounds is 2.1 per cent, despite individual differences as high
as 25 per cent. When an arbitrary division is made at 4 per
cent different, 43 compounds or 15 per cent fall in the deviant
group wherein the difference between calculated and experimental atomization energies exceeds 4 per cent. The average
difference for these compounds is 8.0 per cent. For the other
247 compounds, the average difference is about one per cent,
which seems very satisfactory. Although the average difference of 8 per cent for the 43 deviant compounds is still much
better, to the best of my knowledge, than any results ever
obtained by quantum mechanical approximations, it is still
not acceptable. For further examination, these compounds
are listed in Table 5:14.

Arbitrarily, it was decided that in calculating the
percentage difference between calculated and experimental
atomization energies, where a wide range of possible error
is reported, a calculated value is considered in perfect agreement if it falls within that range and to differ from the nearest
limiting value if it falls outside that range. From Table 5:14
it will be observed that more than half, or 25 compounds,
are normally high melting solids, for which the heats of formation of the gaseous state are presumably obtained with greater
difficulty and perhaps less accurately. For example, poor
agreement is found for all the alkali metal fluoride gases
and for all the gaseous halides of lithium and sodium, but
agreement is good for all the others. The identical parameters
apply here as used in determining the atomization energies
of the solids, which as shown in Chapter 14 all show reasonably
good agreement with experimental values. Similarly, there
are six M2 halides in this table whereas in the solid state
there is relatively little problem in calculating accurate atomization energies. It is not fair to assume without further
evidence that some of the experimental heats of formation
may be in error, but at least it seems possible. On the other
hand, fluorine atoms are the smallest of the halogens and
lithium and sodium atoms the smallest of the alkali metals,

and perhaps there are polarization effects in the gaseous state which cause the atomization energies to be less than calculated. The average difference between calculated and experimental values for the 25 halides under discussion is 11 per cent, which leaves an average of only 5.3 per cent for the remainder.

Three of the remainder are M2 oxides, which in the gaseous state are only singly bonded, and there is some question as to whether normal calculations should suffice. Another half dozen are compounds in which the bond energy is only about 50 kcal (209 kJ) or less, wherein a difference of only 2 kcal (8.4 kJ) makes 4 per cent or more. There are 5 interhalogen molecules included, all of which have experimental bond energies substantially smaller than calculated, and for this no explanation is apparent. In short, this research has not, alas, achieved perfection, or if it has, significant correction in experimental data must be made.

All in all, however, the data provide a remarkable confirmation of the validity of the theory of polar covalence. This confirmation is strongly reinforced by the data in subsequent chapters on organic compounds and highly polar nonmolecular solids. It is to organic chemistry that we next turn attention.

Chapter 6

BONDS AND BOND ENERGIES IN HYDROCARBONS

Standard heats of formation of alkane hydrocarbons vary significantly within a group of isomers by as much as 10 kcal (42 kJ) per mole in nonanes. This fact reveals that the simple theory of polar covalence, which would produce calculated bond energies that are alike for isomers, must be inadequate. Two opposing factors appear to be involved, a strengthening of bonds associated with chain branching, and a weakening of bonds associated with repulsions among branches on adjacent or nearby carbon atoms. Many attempts have been made in the past to develop empirical methods of representing bond energies in branched hydrocarbons and their derivatives (A13, A22-A36), but for the most part these methods have been rather cumbersome and relatively impractical. In this work it was found that fairly accurate atomization energies and heats of formation can be calculated for most organic compounds by use of the simple theory of polar covalence (B4) and ignoring the disturbing factors mentioned above. However, in a search for better understanding and better accuracy, a combination of theory and empiricism has been developed which is more satisfactory in its application to the bonding in a large number of organic compounds representing most common functional groups. More than 750 molecules were studied (A13, A14, B5).

NORMAL ALKANES

First, by assuming bond energies of hydrogen to primary and secondary carbon to be essentially the same, and C'C', C'C", and C"C" bonds also to be essentially equal, it was possible to equate the experimental atomization energies of the normal alkanes through decane with the numbers of each kind of bond. By solving simultaneous equations, CBE's of 82.78 kcal (346.4 kJ) per mole for C-C and 98.82 kcal (413.4 kJ) per mole for C-H were derived. The atomization energies calculated using these bond energies additively are compared with those calculated from experimental heats of formation in Table 6:1. Note that for ethane, in which all bonds involve primary carbon, the calculated atomization energy of 675.6 kcal (2827 kJ) per mole is almost identical with the experimen-

Table 6:1

Atomization Energies and Heats of Formation of n–Alkanes

	Atomization Energy		Heat of Formation	
	calc	exp	calc	exp
C_2	675.64	675.44	-20.44	-20.24±0.12
			4.91	4.59
C_3	956.04	955.52	-25.35	-24.83±0.14
			4.90	5.53
C_4	1236.44	1236.55	-30.25	-30.36±0.16
			4.90	4.74
C_5	1516.84	1516.79	-35.15	-35.10
			4.91	4.82
C_6	1797.24	1797.10	-40.06	-39.92±0.18
			4.90	4.93
C_7	2077.64	2077.53	-44.96	-44.85±0.22
			4.91	5.01
C_8	2358.04	2358.03	-49.87	-49.86±0.25
			4.90	4.80
C_9	2638.44	2638.33	-54.77	-54.66±0.25
			4.90	4.98
C_{10}	2918.84	2918.81	-59.67	-59.64±0.25

Note: In such a series of hydrocarbons, differing by equal increments of CH_2, either equal differences or a steady trend in differences would reasonably be expected. In this light, the calculated heats of formation and atomization energies appear more satisfactory than the experimental values.

tal value of 675.4 kcal (2826 kJ) per mole, justifying the above assumption.

This result can be examined by use of the theory of polar covalence. A methylene group, CH_2, may be considered representative of normal alkanes. Its electronegativity is calculated to be the cube root of C 2.746 x H 2.592 x H 2.592 = 2.642. This corresponds to partial charges of C -0.040 and H 0.020, giving an ionic blending coefficient of 0.030 and a covalent blending coefficient of 0.970. The covalent radius sum is 109.2 pm, and the polarity is too slight to make an appreciable difference in the bond length. The homonuclear energy for hydrogen must be corrected for the small positive partial charge: 104.2 x 0.980 = 102.1 kcal (427.2 kJ), and the geometric mean of this with 82.8 for C is 92.0 kcal (384.9 kJ) per mole. The covalent energy is then 0.97 x 92.0 = 89.2 kcal (373.2 kJ), and the ionic energy is 0.03 x 33200/109.2 = 9.1 kcal (38.1 kJ), giving a bond energy of 98.3, within 0.5 kcal (2.1 kJ) of the experimental value.

Notice that the slight polarity of the C-H bond, only 0.03, nevertheless accounts for more than 9 per cent of the total energy. Polarity always increases the bond strength, and here 2.3 kcal (9.6 kJ) of covalent energy is replaced by 9.1 kcal (38.1 kJ) of ionic energy, strengthening the bond by 6.8 kcal (28.5 kJ).

A feature of organic molecules that differs from most inorganic molecules is the existence of a number, sometimes a fairly large number, of bonds of the same kind. Consequently any small error or discrepancy is multiplied by the number of bonds when the atomization energy is calculated. It has been found that more satisfactory results are obtained using the C-H energy of 98.81 kcal (413.4 kJ) as a standard energy than by attempting to make minor adjustments or corrections.

ALKANE ISOMERS

In all bond energy studies, one must be alert to the possibility of bond weakening by steric effects. Unfortunately these effects cannot usually be quantitatively evaluated, except empirically by comparison of calculated with experimental atomization energies. This is particularly true in branched hydrocarbons and their derivatives. A total of 138 alkane isomers through the decanes were studied. Two effects were noted: (1) a **strengthening** of bonding associated with **chain branching,** and (2) a **weakening** of bonding associated with **proximity of branches.** It was possible to take into account the repulsive effects of branches on adjacent carbon atoms but not very usefully where the branches have a carbon atom between them.

The procedure was to compare similar structures among butanes, pentanes, hexanes, heptanes, octanes, nonanes, and decanes with respect to experimental atomization energies and the standard C-C bond energy found applicable to normal alkanes. By a combination of averaging and solution of simultaneous equations, individual bond energies were obtained

Table 6:2

Hydrocarbon Bond Energies

	C'	C"	C'''	C""	vinyl	C_6H_5
H	98.81	98.81	98.81	98.81	98.81	98.81
C'	82.78	82.78	83.53	84.11		85.8
C"	82.78	82.78	82.94	82.92	84.2	85.8
C'''	83.53	82.94	82.10	80.81		85.4
C""	84.11	82.92	80.81	77.82		84.3

CH_3	296.4	CH_2=CH	440.9	=C-C=	86.7
C_2H_5	576.8	CH_3-CH=CH	723.0	C-C$\equiv$	87.3
nC_3H_7	857.3	C_6H_5	1221.8	=C-C$\equiv$	89.2
iC_3H_7	858.7	C_6H_4	1123.0	$\equiv$C-C$\equiv$	92.4
nC_4H_9	1137.6	$C_6H_5CH_2$	1503.8	C_6H_5-C_6H_5	87.5
sC_4H_9	1138.5	CH_2	280.4	aromatic rings:	
iC_4H_9	1139.2	C=C	144.5	C-C	121.3
tC_4H_9	1141.6	C=C	195.6	C-N	102.8
nC_5H_{11}	1418.0	C-C=	84.4		
nC_6H_{13}	1698.4				

NOTE: These C-C values above account for steric repulsions
        between branches on adjacent carbons but not on carbons
        separated by one. Also, the primes represent the number
        of H atoms attached. Where other atoms are substituted
        for H without causing repulsions, then the correct C-C
        energies to be used are: C'''-C''' 84.28; C'''-C"" 84.86, and
        C""-C"" 85.44 (as in diamond).

as given in Table 6:2, based on the assumption of constant
C-H energy of 98.81 kcal (413.4 kJ). When these values are
assigned in various branched alkanes, they provide agreement
between calculated and experimental atomization energies
within an average of 0.2 kcal (0.8 kJ) for all 123 isomers
in which repulsions occur that are accounted for in the evaluation of the CC single bond energies. Examples are provided
in Table 6:3. In about 15 isomers the repulsions are between
branches on carbon atoms separated by one carbon, and these

Table 6:3

Atomization Energies and Heats of Formation
of Representative iso-Alkanes

Structure	Atomization Energy		Heat of Formation	
	calc	exp	calc	exp
C CCC	1238.69	1238.60	-32.50	-32.41
C CCCC C	1801.37	1801.66	-44.19	-44.48
C C CCCCC	2078.12	2077.93	-45.44	-45.25
C CCCCC C	2081.77	2081.88	-49.09	-49.20
C CCCCCCC	2359.70	2359.64	-51.53	-51.47
CC CCCCC C	2360.77	2360.75	-52.60	-52.58
CC CCCCCCC	2640.17	2640.06	-56.50	-56.39
C C C CCCCC	2640.58	2640.73	-56.91	-57.06
C C C CCCCCCC	2919.98	2920.03	-60.81	-60.86

are not adequately accounted for in the values of Table 6:2.

From the very beginning the usual assignment of a C-C bond energy of about 82-83 kcal (343-7 kJ) has been puzzling because it is smaller than the value of 85.44 kcal (357.5 kJ) for diamond. There is also the question of how or why can chain branching, where there is no reasonable possibility of interchain repulsions, increase the CC bond strength. It now appears that the presence of hydrogen attached to carbon weakens its bonds to other kinds of atoms, or, at least, is associated with such a weakening. Consider, for example, the C'C"" energy which is 1.33 kcal (5.6 kJ) per mole greater than the C'C' energy. If this amount, 1.33 kcal (5.6 kJ) is added to the C'C"" energy, this should now be the C""C"" energy if there were no repulsions between adjacent side chains. The value is now 85.44 kcal (357.5 kJ), essentially that in diamond. The actual C""C"" energy of only 77.82 suggests weakening of 7.62 by repulsions. Similarly, the C'C"' energy is 83.53 kcal (349.5 kJ), which is 0.75 kcal (3.1 kJ) greater than the C'C' value. Adding 0.75 to 83.53 gives 84.28 kcal (352.6 kJ), which should correspond to the CH-CH (or C"'-C"') energy if there were no repulsions between side chains. The actual C"'C"' energy is 82.10, suggesting that the repulsions weaken the bond by about 2.18 kcal (9.1 kJ) per mole of bonds. It seems surprising that there is no significant difference between C'C' and C"C".

CYCLOALKANES

The sum of the two C-H bond energies and one standard C-C bond energy is 280.39 kcal (473.2 kJ) per mole for the methylene group. Multiplying this energy by the number of methylene groups in a cycloalkane gives the expected strain-free atomization energy of the molecule. By subtraction of the experimental atomization energies, this provides a measure of the strain associated with ring closure. The data are presented for a number of cycloalkanes in Table 6:4. As previously well known, only cyclohexane appears to be strain-free.

BENZENE HYDROCARBONS

As described in Chapter 4, multiple bond factors obtained empirically are a linear function of the cube root of the bond order. For benzene, in which the bond order is 1.5, the factor is 1.299. From this and the bond length of 139 pm in benzene, the C-C bond energy is calculated as 121.3 kcal (507.5 kJ) per mole. In earlier work the C-H bond to benzene was believed to be a little weaker than the 98.8 found for alkanes, but there now seems to be less

Table 6:4

Atomization Energies and Strains in Cycloalkanes

cmpd	atomization energy		strain
	calc	exp	
cyclopropane	841.2	813.8	27.4
cyclobutane	1121.6	1095.2	26.4
cyclopentane	1402.0	1395.9	6.1
cyclohexane	1682.4	1682.5	0.0
cycloheptane	1962.8	1956.7	6.1
cyclooctane	2243.2	2233.7	9.5
cyclononane	2523.6	2511.2	12.4

Table 6:5

Atomization Energies and Heats of Formation of Representative Benzene Hydrocarbons

cmpd	atomization energy		heat of formation	
	calc	exp	calc	exp
benzene	1320.4	1320.6	20.0	19.8
toluene	1604.4	1603.9	11.5	12.0
ethylbenzene	1884.9	1884.2	6.5	7.2
i-butylbenzene	2447.9	2447.5	-5.5	-5.1
n-decylbenzene	4128.0	4128.4	-32.6	-33.0
1,2-diphenylethane	3094.0	3095.2	33.6	32.4
1-methyl-4-ethylbenzene				
	2168.4	2167.7	-1.5	-0.8
o-xylene	1888.2	1886.4	2.8	4.6
1,3,5-trimethylbenzene				
	2171.9	2170.7	-5.0	-3.8

Table 6:6

**Atomization Energies and Heats of Formation of
Representative Alkenes**

cmpd	atomization energy		heat of formation	
	calc	exp	calc	exp
1-butene	1102.1	1102.2	-0.1	-0.2
cyclopentadiene	1137.3	1137.2	31.8	31.9
2-methyl-1-pentene	1666.0	1667.2	-13.0	-14.2
2,3-dimethyl-2-butene	1670.7	1669.4	-17.7	-16.4
1,3-cycloheptadiene	1698.1	1697.5	22.0	22.6
5-methyl-1-hexene	1944.9	1944.2	-16.4	-15.7
4,4-dimethyl-1-pentene	1947.4	1947.9	-18.5	-19.0
styrene	1750.4	1751.9	36.8	35.3
1-decene	2784.6	2784.5	-29.6	-29.5

justification for this assumption. The 98.8 value (413.4 kJ) for C-H is also used in determining the atomization energy of benzene. The sum of the contributing bond energies in the phenyl radical is then 1221.8 kcal (5112 kJ) per mole.

From the data of Table 6:2 the atomization energies and heats of formation of various benzene hydrocarbons can be calculated. Representative examples are in Table 6:5. The differences among ortho-, meta-, and para- substitution are not accounted for in this work, but they are relatively small.

ALKENES

The empirical multiplicity factor for the double bond is 1.488 as determined for bond order 2. By use of this and the data of Table 6:2, the standard C=C energy is found to be 144.5 kcal (604.6 kJ). These data can be used additively to obtain reasonably accurate atomization energies and heats of formation of alkenes, as illustrated by the representative data of Table 6:6.

Table 6:7

Atomization Energies and Heats of Formation of Representative Alkynes

cmpd	atomization energy		heat of formation	
	calc	exp	calc	exp
acetylene	393.2	392.6	53.6	54.2
propyne	678.1	678.0	44.2	44.3
2-butyne	963.0	963.1	34.8	34.7
3-methyl-1-butyne	1240.4	1240.7	32.9	32.6
1-hexyne	1519.2	1519.2	29.6	29.6
1-octyne	2080.0	2080.1	19.8	19.7
1-decyne	2640.8	2640.9	10.0	9.9
1-undecyne	2921.2	2921.4	5.1	4.9
1-dodecyne	3201.6	3201.8	0.2	0.0

ALKYNES

Similarly, the empirical factor for the triple bond is 1.787. Application of this and the data of Table 6:2 to determination of atomization energies and heats of formation of a number of representative alkynes is provided in Table 6:7.

OTHER HYDROCARBONS

Condensed ring hydrocarbons and their derivatives have not been included in this study. In addition to whatever special problems these may present, one must in general be alert for the possibility of steric repulsions to reduce the experimental atomization energies below the calculated values.

Chapter 7

SOME ORGANIC HALIDES

First, before considering specific halides, or other functional groups, a general discussion of hydrocarbon derivatives is desirable. In Table 6:2 were given various carbon-carbon bond energies, chosen to minimize problems arising from bond strengthening associated with chain branching and bond weakening associated with steric repulsions among branch chains. For study of the bond energies in organic molecules in general, it is necessary to choose the appropriate homonuclear bond energy of carbon in order to determine its geometric mean with an atom of a functional group. The procedure is, to interpret C'''' as a carbon to which no hydrogen atom is attached, C''' as a CH carbon, C'' as a CH_2 carbon, and C' as a CH_3 carbon. Since in this interpretation a carbon atom does not necessarily possess branch chains substituting for hydrogen, but some functional group atom which may not involve any potential for steric repulsions, it is necessary to use the values derived in Chapter 6 for $C''''C''''$ and $C'''C'''$ without any weakening through steric effects. The appropriate homonuclear energies for carbon are therefore C'''' 85.4 kcal (357.3 kJ) (the same as observed in diamond), C''' 84.3 kcal (352.7 kJ), and 82.8 kcal (346.4 kJ) for C'' and C'. It seems surprising that there is no evidence of a significant difference between the last two, for as will be seen, bonds of methyl to a functional group are nearly always weaker than are bonds of higher normal alkyl groups.

GENERAL PROCEDURE

The general procedure for evaluating bond energy contributions of functional groups is to use the theory of polar covalence to determine contributing bond energies within the functional group and to calculate the approximate energy of the bond connecting the hydrocarbon group to the functional group. The "standard" energy of this bond between hydrocarbon and functional groups is then determined empirically by subtracting the sum of the other energies from the experimental atomization energy of the compound, and averaging the results from as many similar compounds as possible. Unfortunately there is only one methyl derivative of any functional group so no averaging is possible. In compensation, the methyl derivative is likely to be the best known and perhaps its heat of formation has been most accurately determined.

The assumption of constant C-H bond energy is probably inaccurate (A21), but not very significantly so except where there is only one hydrogen atom attached to carbon with highly electronegative substituents, such as HCHO, HCOOH, and so on. In these the C-H bond seems definitely weakened, probably having an energy of around 91-95 kcal per mole instead of 98.8. This is one of the many aspects of application of the theory to organic compounds which **needs further study.**

Finally, it is useful to keep in mind that a distinctive feature of organic chemistry is the occurrence of the same hydrocarbon groups in combination with many different functional groups. The chemistry of these compounds is essentially the chemistry of the functional group, which appears always to be the negative portion of the molecule. In those hydrocarbon derivatives wherein the nonhydrocarbon part is less electronegative than the hydrocarbon part, then it is the hydrocarbon group that is the functional group. For example, organometallic compounds in which the hydrocarbon group is partially negative tend to exhibit similar chemical properties whatever the metal: ease of oxidation, hydrolysis, and so on. The question of how positive or how negative a hydrocarbon group can be without significant changes in the C-C and C-H bond energies **needs further exploration.** And of course there is an inherent conflict of views between those of the organic theorist who wishes to localize a positive or negative charge on a particular atom of a group and those which involve the spreading out of any such charge in effect by partial flow of valence electrons.

Everything considered, it seems rather extraordinary that group and bond energies can be evaluated which are so remarkably additive as to give accurate atomization energies and heats of formation. Indeed they can. In the present study, more than 750 different organic molecules of all common functional types, having an average of 18 bonds per molecule and an average atomization energy of about 1800 kcal (7531 kJ) per mole, are calculated to have atomization energies that differ from the experimental values by an average of only 0.3 per cent.

Development in this field has until recently been rather severely handicapped by censorship. For example, long before I had established the relationship between partial charge and bond energies or even imagined it to be possible, I had calculated partial charges on atoms in organic compounds which were essentially identical to those used so successfully today. Publication of these results in the most appropriate journal was prevented by an anonymous reviewer who stated, "I consider the submission of this manuscript to be an insult to the editor, the referees, and the readers of the Journal." When I queried the editor as to why he should take seriously so intemperate a review, he replied that the referee was of such eminence in his field that his authority could not be questioned. In my opinion, based on many experiences similar to the above, peer review is an indispensable aid to good science publication in areas of general knowledge, but it is capable of gross injustice in the treatment of innovative ideas. This seems most unfortunate because the discovery of new knowledge is the only real excuse for scientific research. New ideas are especially vulnerable during their formation period, urgently needful of tender loving care and very susceptible to the blight of abuse and censorship.

FLUORIDES

As a class, these compounds present considerable difficulty because of the apparent variability of the C–F bond. The bond length is reported to vary from 129 to 143 pm and the energy seems to vary from 106 to 126 kcal (444 to 527 kJ) per mole of bonds. There is no obvious evidence of steric problems but the partial charges are unusually high, in a CF_2 unit being C 0.301 and F –0.150. Internal protonic bridging might well occur under favorable structural conditions, and there is a possibility of repulsions between fluorine atoms on adjacent carbon atoms, resulting from partial charge rather than steric interference. Also, the C–H bond may be affected by fluorine atoms on the same carbon.

By application of the theory of polar covalence, it is calculated that the C–F' bond energy should be in the range of 93.5 to 111.4 kcal (391.2–466.1 kJ) per mole, and that the C–F" energy would range from about 111 to about 128 kcal (391.2–466.1 kJ).

In carbon tetrafluoride, CF_4, wherein the charge on carbon is 0.371 and that on fluorine, -0.092, the ionic energy contribution is calculated to be 58.4 kcal (244.3 kJ), and the covalent contribution is either 49.7 kcal (207.9 kJ) for C-F' or 68.5 (286.6 kJ) for C-F". It is the average of these which is applicable, giving an average bond energy of 117.4 kcal (491.2 kJ) compared with the experimental value of 117.0 kcal (489.5 kJ). This presents one more problem associated with the lone pair bond weakening effect. When the carbon atom is sufficiently positive, does it then have partial use of outer orbitals capable of helping to remove the lone pair bond weakening in the fluorine?

Empirical C-F energies obtained by difference between experimental atomization energies and the sum of the bond energies in the hydrocarbon part are as follows: alkyl C-F: C'F 106.0, C"F 107.6, C'"F 108.3, C""F 113.1; phenyl F 113.2; acyl F 126.2 kcal. These are applicable to monofluorides. In CF_3 attached to nonfluorinated carbon, the C-F energy is 117.5, but when it is part of a perfluorocarbon, it is 115.5 and the CF energy in CF_2 of a perfluorocarbon is 112.1. If it is a perfluoroaryl compound, the energy is 107.0, and perfluoro vinyl is 109 kcal per mole of CF bonds.

Clearly the theoretical values are of the correct order of magnitude but there are too many complications for a complete understanding at present. Table 7:1 gives some representative examples of application of the empirical values together with the data of Table 6:2 to determining the atomization energies and heats of formation of a number of fluorides.

CHLORIDES

The theoretical range for C-Cl' energies is calculated to be 83.3-84.3 kcal (348.5-352.7 kJ) per mole of bonds. The empirical values obtained by difference are as follows: C'Cl 80.9, C"Cl 81.6, C'"Cl 83.7, C""Cl 85.6, vinyl Cl 78.5, and phenyl Cl 83.4. The discrepancies between theory and fact are not large, although they certainly require explanation which is not yet complete.

There is evidence that even two chlorine atoms on the same carbon atom may weaken the bonds, and four certainly do. In CCl_4, the calculated atomization energy is

Table 7:1

**Atomization Energies and Heats of Formation of
Representative Organic Fluorides**

cmpd	atomization energy		heat of formation	
	calc	exp	calc	exp
CF_4	469.6	470.2	-222.7	-223.3
CH_2F_2	421.8	421.4	-108.5	-108.1
CH_3F	402.4	402.4	-55.9	-55.9
nC_3H_7F	964.8	964.7	-67.3	-67.2
C_2F_6	775.8	777.3	-319.8	-321.3
C_7F_{16}	2310.7	2310.8	-808.6	-808.7
C_2F_4	574.9	576.1	-156.7	-157.9
C_6F_6	1369.6	1369.7	-228.4	-228.5
$4-FC_6H_4COOH$	1713.6	1715.9	-115.9	-118.2

346.4 kcal (1449 kJ) compared to an experimental value of only 312.5 kcal (1308 kJ). The bonding appears to be weakened by 34 kcal (142 kJ) per mole by the crowding of the chlorine atoms around the smaller carbon atom. This reveals the fallacy of obtaining a C–Cl standard bond energy by dividing the atomization energy of CCl_4 by 4. In CH_2Cl_2, it is not known whether the C–H bonds are weakened, but if they are not, then the calculated atomization energy is about 12 kcal (50 kJ) greater than the experimental value, suggesting weakening by repulsions. A study of CF_2Cl_2 would remove this uncertainty, but introduce the new uncertainty concerning the amount of reduction in the lone pair bond weakening on fluorine.

Representative applications are presented in Table 7:2. Application to polychlorides is only justifiable where the individual halogen atoms are sufficiently separated in the molecule to avoid repulsions. This is true for all polyhalides beyond the fluorides.

Table 7:2

Atomization Energies and Heats of Formation of
Representative Organic Chlorides

cmpd	atomization energy		heat of formation	
	calc	exp	calc	exp
CH_3Cl	377.3	377.3	-20.6	-20.6
CCl_4	346.4	312.5	-54.7	-25.7
nC_4H_9Cl	1219.2	1218.44	34.4	35.2
$Cl_2C=CH_2$	505.3	504.7	0.9	0.3
$CH_2=CHCl$	519.4	519.4	8.6	8.6
iC_3H_7Cl	942.4	942.7	-34.7	-35.0
2,2-dichloropropane	924.2	926.1	-39.5	-41.4
2-chloro-2-methylbutane				
	1506.3	1507.1	-47.6	-48.4
C_6H_5Cl	1305.2	1305.2	12.2	12.2

BROMIDES

Depending on the range in carbon energy, the C-Br' energy is calculated to be in the range of 69.8-70.7 kcal (292.0-295.8 kJ) per mole of bonds. The empirical values are: C'Br 66.9, C"Br 68.4, C'"Br 69.9, C""Br 70.8, vinyl 66.0, and phenyl 68.0 kcal. The theory is very satisfactory in accounting for the CBr bond energy. Some representative examples of the results of calculating the atomization energies and heats of formation of bromides are given in Table 7:3.

IODIDES

The calculated C-I bond energies range from 57.9 to 58.6 kcal (242.3-245.2 kJ) per mole of bonds. For some reason unknown, perhaps associated with the relatively large size of the iodine atom, the empirical energies are somewhat lower: C'I 53.4, C"I 53.9, C'"I 55.5(?), C""I 55.3, vinyl 54.1, and phenyl, 52.9 kcal. Representative applications are given in Table 7:4.

Table 7:3

**Atomization Energies and Heats of Formation of
Representative Organic Bromides**

cmpd	atomization energy		heat of formation	
	calc	exp	calc	exp
C_2H_5Br	645.2	645.1	−15.4	−15.3
$nC_5H_{11}Br$	1486.3	1487.2	−30.0	−30.9
$nC_6H_{13}Br$	1766.7	1767.7	−34.9	−35.9
$nC_8H_{17}Br$	2327.5	2329.1	−44.7	−46.3
$CH_2=CHCH_2Br$	791.3	789.3	−13.8	−11.8
$CH_2=CHBr$	506.9	506.9	18.7	18.7
$CH_3CHBrCH_3$	928.6	928.5	−23.3	−23.2
$CH_3CHBrCHBrCH_3$	1181.5	1179.8	−26.1	−24.4
C_6H_5Br	1289.8	1289.8	25.2	25.2

Table 7:4

**Atomization Energies and Heats of Formation of
Representative Organic Iodides**

cmpd	atomization energy		heat of formation	
	calc	exp	calc	exp
CH_3I	349.8	349.8	3.3	3.3
C_2H_5I	630.7	630.6	−2.1	−2.0
iC_3H_7I	914.2	914.1	−10.1	−10.0
CH_2ICH_2I	585.8	586.5	16.2	15.5
$CH_3CH=CHI$	777.0	777.0	22.9	22.9
tC_4H_9I	1196.9	1197.2	−17.3	−17.6
$CH_3CH_2CHICH_2I$	1148.2	1150.1	1.0	2.9
C_6H_5I	1274.7	1274.7	39.1	39.1
$C_6H_5CH_2I$	1558.4	1558.7	30.1	30.4

Table 7:5

**Atomization Energies and Heats of Formation of
Representative Acyl Halides**

cmpd	Atomization Energy		Heat of Formation	
	calc	exp	calc	exp
CH_3COF	684.2	683.8	-106.8	-106.4
CH_3COCl	646.8	646.0	-59.2	-58.4
CH_3COBr	630.3	630.8	-45.1	-45.6
CH_3COI	614.7	614.1	-30.9	-30.1
C_6H_5COF	1610.3		-72.2	
C_6H_5COCl	1574.5	1574.4	-26.2	-26.1
C_6H_5COBr	1558.0	1557.5	-12.3	-11.8
C_6H_5COI	1542.4	1542.2	2.3	2.5

ACYL HALIDES

There seems to be some uncertainty about the exact bond length for the CO bond in acyl halides, but from an average value of 122 pm the $C=O''$ energy is calculated to be 177.3 kcal (741.8 kJ) per mole. Assuming this energy to be constant in several acyl halides, the carbon–halogen bonds were found to be represented best as $C-F''$, $C-F'''$ average, $C-Cl''$, $C-Br''$, and $C-I''$, having calculated energies of 126.4, 88.7, 72.7, and 56.4 kcal. These are very close to the empirical energies, which added to the $C=O''$ energy are, for the acyl halide groups, 303.4, 266.0, 250.0, and 234.0 kcal. Table 7:5 gives representative examples of the application of these values. It will be noted (Chapter 8) that single bonds of oxygen to carbonyl carbon, in carboxylic acids and esters, also involve O'' energy, similar to the above halogens.

Chapter 8

ORGANIC COMPOUNDS CONTAINING OXYGEN

ALCOHOLS

The fully weakened oxygen homonuclear bond energy is involved in alcohols. The calculated C-O' energy ranges from about 81 to 82 kcal (339-343 kJ) per mole. The empirical values are, assuming an O-H energy averaging about 112 kcal (469 kJ) per mole, C'O 79.0, C"O 82.0, C'''O 85.4, and C''''O 87.1. The increase is puzzling and will be discussed in more detail at the end of this chapter. It is not the result of a change in oxygen bond energy from O' to O", for the CO" energy is about 111 kcal (464 kJ) per mole. Table 8:1 gives the atomization energies and heats of formation of some representative alcohols. It is suggested that the discrepancy between calculated and experimental values for pentaerythritol may be the result of internal protonic bridging not taken into account in the simple theory.

PHENOLS

Corresponding to a shorter C-O bond length, about 136 compared to 143 pm in alcohols, the CO bond energy is calculated to be 86.0 kcal (359.8 kJ) per mole. The average value for ten phenols determined empirically is 89.9 kcal (376.1 kJ) per mole. Again there at present no explanation for the increase. Atomization energies and heats of formation of a number of representative phenols are given in Table 8:2.

ETHERS

The calculated CO energy range in ethers is the same as for alcohols, but the empirical values are even higher: C'O 83.0, C"O 86.1, C'''O 87.9, C''''O 89.5, vinyl 90.5, and phenyl 92.6 kcal per mole. Here one must expect appreciable repulsions between branch chains on carbons separated by just the one oxygen atom. Table 8:3 gives representative examples in which such repulsions are not expected.

Table 8:1

Atomization Energies and Heats of Formation
of Representative Alcohols

cmpd	atomization energy		heat of formation	
	calc	exp	calc	exp
methanol	487.4	487.4	-48.1	-48.1
ethanol	770.9	771.0	-56.1	-56.2
n-hexanol	1892.4	1892.7	-75.6	-75.9
n-decanol	3014.0	3013.0	-93.2	-94.2
i-propanol	1056.1	1055.4	-65.8	-65.1
t-butanol	1340.7	1340.5	-74.9	-74.7
2-pentanol	1616.4	1616.3	-75.1	-75.0
3-methyl-2-butanol	1617.0	1616.5	-75.7	-75.2
pentaerythritol	1898.2	1905.7	-178.1	-185.6
			(internal protonic bridging?)	

Table 8:2

Atomization Energies and Heats of Formation of
Representative Phenols

cmpd	atomization energy		heat of formation	
	calc	exp	calc	exp
phenol	1423.7	1423.0	-23.7	-23.0
1,2-dihydroxybenzene	1526.8	1524.6	-67.2	-65.0
1,3-dihydroxybenzene	1526.8	1523.2	-67.2	-63.6
1,4-duhydroxybenzene	1526.8	1525.5	-67.2	-65.9
o-cresol	1705.7	1706.2	-30.2	-30.7
p-cresol	1705.7	1705.4	-30.2	-29.9
2,3-dimethylphenol	1987.7	1988.6	-36.7	-37.6
2,6-dimethylphenol	1987.7	1989.7	-36.7	-38.7
4-ethylphenol	1986.1	1985.4	-35.1	-34.4

Table 8:3

Atomization Energies and Heats of Formation of Representative Ethers

cmpd	atomization energy		heat of formation	
	calc	exp	calc	exp
dimethyl ether	758.8	758.8	-44.0	-44.0
methyl ethyl ether	1042.3	1042.0	-52.0	-51.7
methyl butyl ether	1603.1	1603.0	-61.8	-61.7
d iethyl ether	1325.8	1326.1	-60.0	-60.3
e thyl vinyl ether	1194.3	1195.2	-32.7	-33.6
divinyl ether	1062.8	1060.7	-5.4	-3.3
methyl t-butyl ether	1610.5	1610.4	-67.7	-67.6
di-i-propyl ether	1893.2	1893.0	-76.4	-76.2
ethyl phenyl ether	1977.3	1977.3	-26.3	-26.3

ALDEHYDES

There is a small difference between formaldehyde and higher aldehydes with respect to the C=O" energy since there are two hydrogen atoms on the carbon atom in formaldehyde and only one on the aldehyde carbon of other aldehydes. In formaldehyde the C=O" energy is calculated to be 176.1 kcal (736.8 kJ) per mole, whereas for other aldehydes it is 177.0 kcal (740.6 kJ). By difference between the experimental atomization and the calculated CO energy in formaldehyde, a bond energy of 92.5 kcal (387.0 kJ) is found for each C-H bond.   The origin of this weakening is not yet understood, but it seems consistent in formic acid, formaldehyde, hydrogen cyanide, formic esters, and so on. The sum of C-H and C=O" bond energies in the CHO group is therefore 268.6 kcal for formaldehyde and 269.5 for others. Empirically, the average value of 269.2 kcal is obtained. Use of the 269.5 kcal (1128 kJ) value gives, by difference, an average bond energy of C-CHO about 84.4 kcal (353.1 kJ), essentially the same as C-C= where the double bond is to carbon. These values lead to the calculated atomization energies and heats of formation in the representative compounds of Table 8:4.

Table 8:4

**Atomization Energies and Heats of Formation of
Representative Aldehydes**

cmpd	atomization energy		heat of formation	
	calc	exp	calc	exp
formaldehyde	361.1	361.1	-26.0	-26.0
acetaldehyde	650.3	650.3	-39.7	-39.7
propanal	930.7	931.6	-44.6	-45.5
butanal	1211.1	1210.5	-49.5	-48.9
heptanal	2052.3	2051.2	-64.2	-63.1
2-methylpropanal	1212.8	1213.9	-51.2	-52.3
2-ethyl-1-hexanal	2334.8	2335.2	-71.2	-71.6
benzaldehyde	1580.1	1580.1	-8.8	-8.8

KETONES

Here a typical calculated C=O" energy is 177.9 kcal (744.3 kJ) per mole. There is not much difference among the empirical C-CO energies obtained by difference, for the different alkyl groups, C' and C" being about 83.8, C'" 83.5, and C"" averaging about 82.7 kcal per mole. For alkyl phenyl ketones the phenyl CO bond energy averages 88.0 kcal (368.2 kJ), but in diphenyl ketone it is only 85.6 kcal (358.2 kJ), suggesting some repulsion between the phenyl groups. Di-t-butyl ketone also appears to show some repulsion amounting to about 5 kcal per mole. Representative examples are supplied in Table 8:5.

CARBOXYLIC ACIDS

In carboxylic acids, an average O-H energy of about 114 kcal (477 kJ) per mole is calculated, and the C=O" energy is about 176 kcal (736 kJ). For the single bond, C-OH, the CO' energy averaged with the CO" energy is about 101 kcal per mole, which corresponds to the empirical average found by difference for eleven carboxylic acid molecules. Thus the bond of carbonyl carbon to hydroxyl oxygen appears to have part of the LPBWE removed. The sum of the calculated averages for C=O, C-OH, and O-H is 391 kcal (1636 kJ), which

Table 8:5

Atomization Energies and Heats of Formation of Representative Ketones

cmpd	atomization energy		heat of formation	
	calc	exp	calc	exp
acetone	938.3	938.0	-52.2	-51.9
methyl ethyl ketone	1218.7	1218.6	-57.1	-57.0
diethyl ketone	1499.1	1498.8	-62.0	-61.7
methyl isopropyl ketone	1500.3	1499.9	-63.2	-62.8
methyl t-butyl ketone	1782.4	1781.9	-69.8	-69.3
ketene	520.0	520.6	-13.6	-14.2
propyl phenyl ketone	2428.7	2428.4	-30.9	-30.6
di-t-butyl ketone	2626.5	2621.7	-87.4	-82.6*
diphenyl ketone	2797.5	2792.8	9.9	14.6*

*steric weakening

Table 8:6

Atomization Energies and Heats of Formation of Representative Carboxylic Acids

cmpd	atomization energy		heat of formation	
	calc	exp	calc	exp
CH_3COOH	773.5	773.5	-103.3	-103.3
nC_3H_7COOH	1334.3	1333.6	-113.1	-112.4
$nC_8H_{17}COOH$	2736.3	2736.7	-137.6	-138.0
$nC_{18}H_{37}COOH$	5540.0	5541.4	-186.3	-187.7
$(COOH)_2$	859.8	860.2	-174.6	-175.0
maleic acid	1296.3	1294.4	-164.3	-162.4
adipic acid	1993.0	1994.0	-205.8	-206.8
sebacic acid	3114.6	3109.6	-225.4	-220.4
benzoic acid	1699.6	1701.0	-68.7	-70.1

Table 8:7

**Atomization Energies and Heats of Formation of
Representative Esters**

cmpd	atomization energy		heat of formation	
	calc	exp	calc	exp
CH_3COOCH_3	1043.8	1043.6	−98.1	−97.9
$CH_3COOC_2H_5$	1327.6	1327.5	−106.4	−106.3
$CH_3COOC_4H_9$	1888.4	1888.8	−115.7	−116.1
$C_4H_9COOCH_3$	1885.0	1884.9	−112.8	−112.7
$C_4H_9COOC_2H_5$	2168.8	2168.9	−121.1	−121.2
$C_7H_{15}COOCH_3$	2726.2	2726.2	−127.5	−127.5
$CH_3COOCH(CH_3)_2$	1612.0	1611.8	−115.3	−115.1
$C_4H_9COOCH(CH_3)_2$	2453.2	2453.4	−130.0	−130.2
$C_6H_4(COOC_2H_5)_2$	3186.7	3187.9	−163.3	−164.5

is taken as the standard COOH energy. An empirical average of 86.1 kcal (360.2 kJ) is then found for the C-COOH bond. Data for branched alkyl carboxylic acids are not available. Representative applications are provided in Table 8:6.

CARBOXYLIC ESTERS

Here there are a number of uncertainties requiring arbitrary selection. If the same energy is assumed for esters as for acids, in the R-COO bond, 86.1 kcal (360.2 kJ), and the C=O″ average is calculated as 178.8 kcal (748.1 kJ), per mole, and the C-OR bond is calculated as a C-O″ which differs from the C-OH bond, having an energy of 107.8 kcal (451.0 kJ), then a sum for the ester group is 372.6 kcal (1559 kJ) per mole, including the R-C bond. Using this and other appropriate data from Table 6:2 for determining the sum, and subtracting from the experimental atomization energy to get the difference, the ether type C-O bond energies have the following empirical values: O-C′ 78.3, O-C″ 81.7, O-C‴ 84.2 kcal, resembling the ethers. On the other hand, if the carbonyl C-O single bond is, like that in carboxylic acids, an average of C-O' and C-O", this means that about 7 kcal per mole must be added to the ether-type bond energies just given. Table 8:7 gives representative results, rejecting the last suggestion which seems less likely.

DISCUSSION

As will be summarized at the end of these chapters on organic compounds, the "strengthening" of the carbon bonds associated with chain branching appears to be a result, instead, of **weakening** the normal carbon bond energy by the attachment of hydrogen atoms to carbon. This allows us no explanation, however, of the carbon-oxygen bond in ethers. In highly branched ethers, assuming no repulsions among branches, the CO bond appears to be **stronger** than calculated. It would seem that the calculated value should represent an upper limit and not be exceeded by the experimental value. The hydrogen atoms bear positive charge but it is quite small. Perhaps branching on the carbon next to other oxygen permits internal protonic bridging which adds to the total atomization energy, but this is an explanation offered in the total absence of other conceivable possibilities.

Chapter 9

ORGANIC COMPOUNDS CONTAINING SULFUR

MERCAPTANS--THIOLS

The theory of polar covalence gives a calculated S-H bond energy averaging 88.6 kcal (370.7 kJ) per mole, and at a bond length of 162 pm, the CS bond energy ranges from 72 to 73,1 kcal (301-305 kJ) per mole. This is for the fully weakened bond, C-S'. By difference between experimental atomization energy and the sum of the contributing bond energies in the hydrocarbon group and the S-H group, the following empirical CS bond energies are obtained: C'S 66.7, C"S 67.6, C'"S 68.5, and 69.7 kcal for phenyl S. Representative examples are presented in Table 9:1.

SULFIDES

The same CS bond energy is calculated for sulfides as for thiols, ranging from 72 to 73.1 kcal (301-305 kJ) per mole. Empirically by difference the following values were obtained: C'S 69.0, C"S 69.6, and phenyl 71.7 kcal. Data for more branched hydrocarbon groups are not available. For disulfides, RSSR, assumption of the same CS energies as in sulfides leaves an average S-S bond energy of 63.9 kcal (267.4 kJ) per mole. The homonuclear bond energies for sulfur are, S' 54.9, S" 60.4, and S'" 65.9 kcal. Thus the 63.9 value is roughly an average of S"-S" and S'"-S'". This corresponds closely to the 63.5 observed experimentally for the S-S bond in gaseous S_8. If the S"-S" bond energy is applicable, then the CS energy averages 71.1 kcal (297.5 kJ) per mole, and if the S'-S' is correct for this application, the CS average is 73.8 kcal (303.8 kJ) per mole.

Table 9:2 provides representative examples. For the disulfides in the table, it is assumed that the CS bond is the same as in thiols and the S-S energy is 63.9 kcal. The good agreement does not guarantee that the correct choice has been made.

SULFOXIDES

The S-O bond here is assumed to be S-O", for which the calculated bond energy is 92.7 kcal (387.9 kJ) per mole.

Table 9:1

Atomization Energies and Heats of Formation of Representative Thiols

cmpd	atomization energy		heat of formation	
	calc	exp	calc	exp
CH_3SH	451.7	451.7	-5.4	-5.4
C_2H_5SH	733.0	732.8	-11.2	-11.0
C_3H_7SH	1013.5	1013.4	-16.2	-16.1
iC_3H_7SH	1015.8	1015.4	-18.5	-18.1
nC_4H_9SH	1293.8	1293.8	-21.0	-21.0
sC_4H_9SH	1295.6	1295.9	-22.8	-23.1
$nC_5H_{11}SH$	1574.2	1574.5	-25.9	-26.2
$nC_6H_{13}SH$	1854.6	1854.7	-30.8	-30.9

Table 9:2

Atomization Energies and Heats of Formation of Representative Sulfides

cmpd	atomization energy		heat of formation	
	calc	exp	calc	exp
CH_3SCH_3	730.8	730.7	-9.0	-8.9
CH_3SSCH_3	795.4	794.2	-7.0	-5.8
$CH_3SC_2H_5$	1011.8	1011.5	-14.5	-14.2
$C_2H_5SSC_2H_5$	1356.2	1357.2	-16.8	-17.8
$CH_3SC_3H_7$	1292.3	1292.4	-19.5	-19.6
$CH_3SC_4H_9$	1572.6	1572.7	-24.3	-24.4
$CH_3SC_6H_5$	1658.9	1658.9	23.6	23.6

Table 9:3

Atomization Energies and Heats of Formation of Representative Sulfoxides

cmpd	atomization energy		heat of formation	
	calc	exp	calc	exp
CH_3SOCH_3	817.5	817.5	-36.1	-36.1
$(C_2H_5)_2SO$	1382.5	1381.5	-50.1	-49.1
$(C_3H_7)_2SO$	1943.5	1944.3	-60.1	-60.9
$C_2H_5SOCH_2CH=CH_2$	1528.7	1528.4	-25.0	-24.7
$(C_6H_5)_2SO$	2677.2	2677.2	25.6	25.6

For the CS bond the energy is calculated to range from about 72 to 73 kcal (301-305 kJ) per mole. The empirical values, however, are C'S 66.0, C"S 68.1, and phenyl S 70.5, data being unavailable to determine the other possibilities. Examples are provided in Table 9:3.

SULFONES

Here it is assumed that the SO bonds are an average of S-O" and S-O''', which is calculated as 100.4 kcal (420.1 kJ) per mole. The empirical CS values obtained assuming this is the correct SO value are: C'S 68.3, C"S 70.9, C'''S 71.5, vinyl-S 67.7, and 73.7 for phenyl-S. Examples of the appllication of these are given in Table 9:4. Alternative selections of bond type for the SO bonds give less reasonable CS energies.

SULFITES AND SULFATES

If it is assumed that the SO bond lengths in organic sulfites are the same as in sulfates, 153 pm for HO-S and 142 for SO, then the calculated bond energies, assuming S-O" bonds, are 86.8 and 93.5 kcal, for a total of 267.1 kcal (1117.5 kJ) per sulfite group. This gives, by difference, empirical CO bond energies of C'O' 78.9, C"O' 80.6 kcal per mole. Data for five sulfites are given in Table 9:5.

For sulfates the SO_4 bond energies add up to 360.8

Table 9:4
**Atomization Energies and Heats of Formation of
Representative Sulfones**

cmpd	atomization energy		heat of formation	
	calc	exp	calc	exp
$(CH_3)_2SO_2$	930.1	930.1	-89.1	-89.1
$CH_3SO_2C_2H_5$	1213.1	1214.0	-96.7	-97.6
$CH_3SO_2CH(CH_3)_2$	1495.6	1495.6	-103.6	-103.6
$(CH_2=CH)_2SO_2$	1217.9	1219.6	-34.3	-36.0
$C_2H_5SO_2CH_2CH=CH_2$	1642.3	1640.3	-79.0	-77.0
$(nC_4H_9)_2SO_2$	2617.7	2615.8	-123.7	-121.8
$C_6H_5SO_2CH=CH_2$	2004.8	2003.8	-31.8	-30.8
$(C_6H_5)_2SO_2$	2791.7	2790.7	-29.3	-28.3
$(C_6H_5CH_2)_2SO_2$	3350.1	3350.9	-36.7	-37.5

Table 9:5
**Atomization Energies and Heats of Formation
Representative Organic Sulfites and Sulfates**

cmpd	atomization energy		heat of formation	
	calc	exp	calc	exp
$(CH_3)_2SO_3$	1017.7	1015.6	-117.6	-115.5
$(C_2H_5)_2SO_3$	1581.9	1583.0	-130.8	-131.9
$CH_3SO_3C_2H_5$	1299.8	1300.8	-124.2	-125.2
$(nC_3H_7)_2SO_3$	2142.9	2142.6	-140.8	-140.5
$(nC_4H_9)_2SO_3$	2703.5	2702.5	-150.4	-149.4
$(CH_3)_2SO_4$	1123.8	1123.7	-164.2	-164.1
$(C_2H_5)_2SO_4$	1690.0	1691.3	-179.4	-180.7
$(nC_3H_7)_2SO_4$	2251.0	2250.8	-189.5	-189.2
$(nC_4H_9)_2SO_4$	2811.6	2810.6	-199.0	-198.0

(1509.6 kJ) per mole, there being two bonds of each length.
From this, empirical CO bond energies are obtained by difference as follows: C'-O' 85.1, and C"-O' 87.8 kcal. Data for
four sulfates are also included in Table 9:5. Selection of
the S-O" bond type for these compounds agrees with H_2SO_4.

Chapter 10

SOME ORGANIC NITROGEN COMPOUNDS

NITRILES

There has been no reason for considering the CN bond in nitriles to be other than a normal triple bond, for which the calculated bond energy is 233 kcal (975 kJ) per mole. If this is subtracted from the atomization energy of HCN determined experimentally, 304 kcal (1272 kJ) per mole, only 71 kcal (297.1 kJ) is left for the H-C bond. This bond would be expected to be more like that in aldehydes and formic acid, about 92–94 kcal (385–393 kJ) per mole. If the bond energies in CH_3 and the calculated CN value of 233 are subtracted from the atomization energy of acetonitrile, CH_3CN, which is 590.9 kcal (2472.3 kJ), this leaves only 62.5 kcal (261.5 kJ) for the C-CN bond. On the other hand, if the C-CN energy is assumed to be the same as in alkynes for the C-C$\equiv$ (the experimental bond lengths are indeed essentially the same), which is 87.4, the CN energy by difference is only 207.2 kcal (886.9 kJ) per mole. Assuming the C-CN energy to be 87.4, the CN energy by difference becomes 211.1 for propionitrile and 210.1 for butyronitrile, averaging 210.6 kcal (881.2 kJ) per mole. This same value in acetonitrile would require the C-CN energy to be 83.9 instead of 87.3 kcal per mole. If the CN bond were truly a triple bond, of the normal variety, then this would correspond to a triple bond factor of 1.616 instead of 1.787. The average of the double bond factor, 1.488, and the triple bond factor 1.787, is 1.638. Clearly something unusual is indicated. The CN bond does not appear to be a normal triple bond.

The energy calculated for the $C=N'''$ bond is 194.1 kcal (812.1 kJ) per mole. Averaged with the triple bond energy of 233.0 this gives 213.6 kcal (893.7 kJ) for the bond energy. The difficulty arises in the presence of two dependent variables, the CN energy and the C-CN energy, but there seems little doubt that the CN bond is intermediate between a double and triple bond. Perhaps this should not be surprising, in view of the difference between acetylene and the C_2 molecule, and in comparison with N_2. A triple bond between carbon atoms appears to require that the fourth bonding orbital of each carbon atom be occupied in a bond, either to another

Table 10:1

**Atomization Energies and Heats of Formation of
Representative Nitriles**

cmpd	atomization energy		heat of formation	
	calc	exp	calc	exp
HCN	304.1	304.1	32.3	32.3
CH_3CN	590.9	590.9	21.0	21.0
C_2H_5CN	874.8	875.3	12.6	12.1
C_3H_7CN	1155.3	1154.8	7.6	8.1
$CH_2=CHCN$	739.1	739.1	44.1	44.1
$NC(CH_2)_3CN$	1354.4	1354.2	40.7	40.9
C_6H_5CN	1521.0	1521.1	51.6	51.5
iC_3H_7CN	1156.7	1156.7	6.2	6.2

carbon or to hydrogen. This observation is based on the fact that when both hydrogen atoms are removed from acetylene, the bond apparently changes from triple to double, for the experimental evidence supports a double bond in C_2. If a C=C bond and an N≡N bond are averaged, then an intermediate CN bond is expected. However, the fourth bonding orbital on carbon in nitriles is indeed occupied, so the puzzle has not been explained at all.

Examination of the cyanogen molecule, C_2N_2, is not very helpful. The experimental C-C bond length is 138 pm, from which the bond energy is calculated to be 95.3 kcal (398.7 kJ). This, subtracted from the experimental atomization energy of 494.8 kcal (2070 kJ), allows only 199.7 kcal (835.5 kJ) for the CN bond energy.

By adopting somewhat arbitrarily the following empirical C-CN energies, C'-CN 83.8, C"-CN 87.3, phenyl-CN 89.2 kcal per mole, the atomization energies and heats of formation presented in Table 10:1 were obtained.

AMINES

A typical N-H bond is calculated to have an energy close to 94 kcal (393 kJ) per mole. From this and the other

Table 10:2

Atomization Energies and Heats of Formation of Representative Amines

cmpd	atomization energy		heat of formation	
	calc	exp	calc	exp
CH_3NH_2	550.1	550.3	-5.3	-5.5
$(CH_3)_2NH$	824.8	824.7	-4.5	-4.4
$(CH_3)_3N$	1101.6	1101.5	-6.0	-5.9
$C_2H_5NH_2$	832.3	831.7	-12.0	-11.4
$(C_2H_5)_2NH$	1389.0	1388.5	-17.7	-17.2
$(C_2H_5)_3N$	1944.3	1944.4	-22.0	-22.1
$iC_3H_7NH_2$	1115.8	1115.8	-20.0	-20.0
$tC_4H_9NH_2$	1400.2	1400.2	-28.9	-28.9
$C_6H_5NH_2$	1484.7	1484.7	20.8	20.8

bond energies, subtracting their sum from the experimental atomization energy, the following values for C-N energies were found for primary amines: C' 65.7, C" 67.5, C'" 69.1, C"" 70.6, and phenyl 74.9 kcal. In all, the fully weakened bond energy of nitrogen is involved. For secondary amines, the C-N energies are, C' 69.0, C" 70.7, C'" 72.7, and phenyl 75.0 kcal. For tertiary amines, these energies are C' 70.8, C" 71.3, and data are not available for determining the others. Note that the energies are weaker the more hydrogen atoms on the carbon and the more hydrogen atoms on the nitrogen. This will be discussed later.

Representative amines with their atomization energies and heats of formation are presented in Table 10:2.

AMIDES

The bond energies within the amide group, $CONH_2$, are calculated to have an average value of 450.5 kcal (1884.9 kJ): C=O" 178.8, C-N' 86, N-H 92.7. The CN value of 86 is an average of C-N' and C-N" calculated. The C-$CONH_2$ bond is taken to be the same as in acids and esters, 86.1 kcal

Table 10:3

Atomization Energies and Heats of Formation of Representative Amides

cmpd	atomization energy		heat of formation	
	calc	exp	calc	exp
CH_3CONH_2	832.7	832.8	-56.8	-56.9
$C_2H_5CONH_2$	1113.1	1113.1	-61.9	-61.9
$C_3H_7CONH_2$	1393.6	1393.4	-66.9	-66.9
$C_4H_9CONH_2$	1673.9	1671.5	71.8	69.4
$C_5H_{11}CONH_2$	1954.3	1956.5	-76.6	-78.8
$C_7H_{15}CONH_2$	2515.1	2515.4	-86.4	-86.7
NH_2CONH_2	723.3	724.0	-58.0	-58.7

(360.2 kJ) per mole. From these and the appropriate hydrocarbon values, the atomization energies and heats of formation of representative amides given in Table 10:3 were found.

NITRO COMPOUNDS AND NITRATES

For the nitro group, with an NO bond length of 122 pm, an N"=O" bond is calculated to have an energy of 139.5 kcal (583.7 kJ), and an N'-O" bond an energy of 76.3 kcal (319.2 kJ) per mole. Their average represents the actual bonds, so the sum of bond energies in the NO_2 group is their sum, 215.8 kcal (902.9 kJ) per mole. From this and the appropriate hydrocarbon bond energies, the following empirical C-N' energies are found: C'N 65.5, C"N 67.4, C'''N 69.8, and phenyl 68.2 kcal per mole. These may be compared with the calculated value for a typical C-N' bond, about 71.5 kcal (299.2 kJ). Table 10:4 gives some representative data for nitro compounds.

In organic nitrates, the carbon is joined to oxygen instead of nitrogen. Within the nitrate group there are two bonds like those in the nitro group, having the calculated energy of 214.8 kcal (902.9 kJ), and one N'-O' bond to the oxygen bridging the nitrogen and carbon, for which the energy is calculated to be 53.1 kcal (222.2 kJ) per mole. The sum

Table 10:4

**Atomization Energies and Heats of Formation of
Representative Nitro Compounds and Nitrates**

cmpd	atomization energy		heat of formation	
	calc	exp	calc	exp
CH_3NO_2	577.7	577.7	-17.9	-17.9
$C_2H_5NO_2$	860.0	859.7	-24.7	-24.4
$iC_3H_7NO_2$	1144.3	1144.0	-33.5	-33.2
$nC_4H_9NO_2$	1420.4	1420.7	-34.1	-34.4
$pNO_2C_6H_4NH_2$	1669.9	1669.4	15.7	16.2
CH_3ONO_2	648.5	648.5	-29.1	-29.1
$C_2H_5ONO_2$	931.6	931.7	-36.7	-36.8
$nC_3H_7ONO_2$	1212.1	1212.0	-41.8	-41.7

for the nitrate group is 268.9 kcal (1125 kJ). By difference the empirical values of C'O 83.2 and C"O 85.9 kcal per mole are obtained. Available data on nitrates are also included in Table 10:4.

SUMMARY OF ORGANIC APPLICATIONS

Data for the calculated bond energies within functional groups are summarized in Table 8:5. The empirical energies for the attachment of carbon to the functional groups are summarized in Table 8:6. Applied together with the data of Table 6:2 which supplies the bond energies in hydrocarbon groups, these allow the calculation of atomization energies of the more than 750 compounds studied within an average of about 0.3 per cent of the experimental values. The tables in Chapters 6-10 provide 228 representative examples of these. Since the average number of bonds in these molecules is 18 and the average atomization energy 1800 kcal (7531 kJ), the 0.3 per cent amounts to about 5 kcal (21 kJ) per mole. If this discrepancy is spread out through the 18 bonds it is quite negligible, but if it is present in one bond, then improvement is still needed. In general, the calculated bond energies in and to the functional groups are quite satisfactorily consistent with the total atomization energies, calculated and

Table 10:5

Bond Energies within Functional Groups
Calculated from Theory of Polar Covalence

-COF	303.7	(C=O^{ii} 177.3; C-F",F' 126.4)
-COCl	266.0	(C=O$^{::}$ 177.3; C-Cl" 88.7)
-COBr	250.0	(C=O$^{::}$ 177.3; C-Br" 72.7)
-COI	233.7	(C=O$^{:1}$ 177.3; C-I" 56.4)
-OH	112.0 aver,	(111.7-112.2)
-CHO	269.5	(C=O$^{:}$177.0; C-H 92.5 by difference)
-CO-	177.9	(C=O)
-COOH	391.0	(O-H 114.0; C=O 176.0; C-O',O" 101.0)
-COOR	286.6	(C=O$^{:}$ 178.8; C-O" 107.8)
-CONH$_2$	450.2	(C=O^{ii} 178.8, C-N' 86.0; N-H 92.7)
-NH$_2$	188.0	(N-H 94.0)
-CN	210.7 emp. value (aver C=N"', C=N"'	
-NO$_2$	215.8	(aver N"=O", N'-O')
-ONO$_2$	268.9	(N'-O' 53.1, NO$_2$ 215.8)
-SH	88.6 aver	
-SO-	92.7	
-SO$_2$-	200.7	(aver S-O" 92.7 and S-O"' 108.0)
-SO$_3$-	267.1	
-SO$_4$-	360.8	(2 x 86.8 + 2 x 93.6)

experimental. In this respect it may be said that the theory of polar covalence, especially when improved by taking into account relatively minor but still significant factors largely concentrated in organic molecules, gives quite an accurate account of the **bond strengths,** and therefore the **atomization energies,** and from these the **heats of formation,** of an extraordinary number and variety of organic molecules as well as the inorganic compounds previously discussed. In other words, there is now available to us a depth of understanding of chemistry which can be very helpful in assisting our chemical intuition and in show us more clearly than previously how the individual atoms contribute to the properties of molecules.

Table 10:6

Empirical Energies for Bonds of Carbon to Functional Group

	C'	C"	C'"	C""	vinyl	C_6H_5	calc range
mono-F	106.0	107.6	108.3	113.1	(102.)	113.2?	
in CF$_3$ of perfluoro				115.5			
in CF$_2$ of perfluoro				112.1			111-128
perfluoro-aryl				107.0			
CF$_3$ to nonfluoro				117.5			
in CF$_4$				117.4			
mono-Cl	80.9	81.6	83.7	85.6	78.5	83.4	83.3-84.3
mono-Br	66.9	68.4	69.9	70.8	66.0	68.0	69.8-70.7
mono-I	53.4	53.9	55.	55.3	54.1	52.9	57.9-58.6
-OH	79.0	82.0	85.4	87.1		89.9	81.0-81.8
-OR	83.0	86.1	87.9	89.5	90.5	92.6	81.0-81.8
-CHO	84.4	84.4	86.1	87.	86.9	88.8	84.4
-COR	83.8	83.8	83.5	82.7	86.7	88.0	84.4
-COOH	86.1	86.1	85.3	86.0	88.3	86.7	86.7
COO-R	78.3	81.7	84.2			86.7	86.7
-CONH$_2$	86.1	86.1					86.1
-NH$_2$	65.7	67.5	69.1	70.6		74.9	71.5-72.4
-NHR	69.0	70.7	72.7	73.8	74.0	75.0	71.5-72.4
-NR$_2$	70.8	71.3					71.5-72.4
-CN	83.8	83.8	83.8	85.7	87.5	88.5	87.3?
-NO$_2$	65.5	67.4	69.8	71.1	69.2	68.2	71.5-72.4
-ONO$_2$	83.2	85.9					81-81.8
-SH	66.7	67.6	68.5	70.5		69.7	72-73.1
-SR	69.0	69.6	70.4	71.2	71.9	71.7	72-73.1
-SOR	66.0	68.1				70.5	72-73.1
-SO$_2$R	68.3	70.9	70.8	72.8	67.7	73.7	72-73.1
-OSO$_2$R	78.9	80.6					
-OSO$_3$R	85.1	87.8					

There remain, of course, many areas deserving further study. Particularly intriguing is the problem of explaining the differences between the different bond energies, those that appear to be influenced by the number of hydrogen atoms attached to carbon or other atom. A detailed study of Table 10:6 will reveal that there is an upward trend in energy from C' through C"", in other words, from bonds to CH_3 to bonds to C, in fluorides, chlorides, bromides, iodides, alcohols, ethers, aldehydes, (but not ketones or carboxylic acids), esters, amines, (probably not in nitriles), nitro compounds and possibly nitrates, thiols, sulfides, possibly sulfoxides, and sulfones. It will also be noted that tertiary amines appear to have stronger bonding to hydrocarbon groups than secondary amines, which in turn have stronger bonds than in primary amines. Furthermore, the bonding appears to be stronger in ethers than in alcohols, and in sulfides than in thiols. These observations suggest that the attachment of hydrogen atoms to other atoms, here carbon, oxygen, nitrogen, and sulfur, may weaken the bonds which the other atom can form elsewhere. They are consistent with the data of Table 6:2 if allowance is made for weakening by steric effects. The mechanism of this weakening would indeed be interesting to learn. There seems to be little evidence of this in inorganic compounds of hydrogen.

Another problem exposed by the data of Table 10:6 is that although the most common trend is toward the calculated bond energy as an upper limit, which is consistent with the apparent hydrogen weakening effect, this is not true for alcohols and ethers, at least, and possibly for aldehydes, wherein the empirical energies exceed the calculated values by a substantial amount. **Explanation is certainly needed.**

To determine the atomization energy of an organic compound, add the hydrocarbon energy from Table 6:2, p 94, to the energy of carbon to functional group, Table 10:6, p 127, and the functional group energy, Table 10:5, p 126. **To determine the heat of formation,** add the atomization energies of all the component atoms, Table 2:1, p 31, and from this sum, subtract the atomization energy of the compound.

Chapter 11

BOND ENERGIES IN ORGANOMETALLIC COMPOUNDS

Heats of formation of organometallic compounds are, so far, especially difficult to measure. The usual combustion calorimetry produces results not easy to interpret, because the combustion products are frequently not clean cut. For example, the organic portion may be converted reasonably well to carbon dioxide and water, but the metal may become a mixture of carbide, carbonate, oxides, and even nitrides, nitrites, or nitrates, as well as free metal. The true heat of formation cannot easily be determined. These and associated problems have been considered by Skinner (A43) and by Steele (A42), as well as by Cox and Pilcher (A13).

As a consequence of such problems, a thorough quantitative treatment of the bond energies of organometallic compounds, and organononmetallic as well, is not yet possible. However, sufficient data are available to reveal an unexplained phenomenon, the fact that the theory of polar covalence yields energies of the bonds of the central atom to the hydrocarbon group that are substantially higher than those determined by difference.(B4, 1976, p 189).

In determining the "experimental" M-C bond energy by difference, the calculated energies of the other bonds are subtracted from the experimental atomization energy. The assumption is made that the bond energies within the hydrocarbon group are those selected as "standard," in Table 6:2, p 91. For a few compounds the M-C bond lengths were not available and required estimation. The bond energy data for alkyls are given in Table 11:1, and for phenyl compounds in Table 11:2. The difference between the calculated M-C and the "experimental" value is considered to be a measure of weakening (W) of the M-C bond.

The arbitrary nature of this assignment must be recognized at once. The only fact known for certain is that the calculated atomization energy of the gaseous compound does differ from the experimental value, being normally larger by a substantial amount. [The only exception noted thus far is boron triphenyl in Table 11:2. Boron compounds are

Table 11:1

Bond Energy Data for Some Alkyl Compounds of the Elements

cmpd	charge E	C	R_o	E_c	E_i	Ecalc	$\Delta Hf°$	C-E	Wkng	%W
ZnM_2	0.153	-0.065	(185.)	54.1	19.5	73.6	13.1	40.3	33.3	45.2
ZnE_2	0.163	-0.054	(185.)	54.1	19.5	73.6	13.6	35.1	38.5	52.3
ZnP_2	0.169	-0.048	(185.)	54.1	19.5	73.6	-2.9	38.5	35.1	47.7
ZnB_2	0.170	-0.048	(185.)	54.1	19.5	73.6	-11.9	38.1	35.5	48.2
CdM_2	0.258	-0.076	(200.)	47.3	27.7	75.0	25.8	31.7	43.3	57.7
CdE_2	0.275	-0.062	(200.)	47.2	28.1	75.3	25.5	26.9	48.4	64.3
HgM_2	0.164	-0.065	207.	25.9	18.4	44.3	22.3	27.4	16.9	38.1
HgE_2	0.176	-0.055	207.	25.9	18.6	44.5	17.8	24.8	19.7	44.3
HgP_2	0.180	-0.051	207.	25.9	18.6	44.5	8.2	24.7	19.8	44.5
HgB_2	0.183	-0.048	207.	25.9	18.6	44.5	-7.8	27.8	16.7	37.5
BM_3	0.138	-0.049	156.	73.5	20.6	94.1	-29.3	87.3	6.8	7.2
BE_3	0.145	-0.049	156.	73.5	20.6	94.1	-36.5	84.8	9.3	9.9
BP_3	0.148	-0.047	156.	73.4	20.9	94.3	-56.4	86.5	7.8	8.3
BB_3	0.149	-0.045	156.	73.5	20.6	94.1	-68.6	80.4	13.7	14.6
AlM_3	0.404	-0.078	(195.)	49.9	41.0	90.9	-20.9	64.4	26.5	29.2
AlE_3	0.423	-0.062	(195.)	49.8	41.4	91.2	-39.1	65.6	25.6	28.1
GaM_3	0.079	-0.051	198.	55.4	10.9	66.3	-11.2	56.7	9.6	14.5
GaE_3	0.084	-0.047	198.	55.4	10.9	66.3	-17.9	54.1	12.2	18.4
GaB_3	0.088	-0.043	198.	55.4	10.9	66.3	-53.1	56.0	10.3	15.5
InM_3	0.196	-0.061	(215.)	47.3	19.9	67.2	40.8	36.9	30.3	45.1
SiM_4	0.200	-0.057	187.	60.5	22.9	83.4	-57.1	72.4	11.0	13.2
SiE_4	0.208	-0.050	187.	60.5	22.9	83.4	-64.4	69.3	14.1	16.9
GeE_4	0.006	-0.043	195.	60.8	4.3	65.1	-39.6	58.6	6.5	10.0
GeP_4	0.007	-0.042	195.	60.8	4.3	65.1	-54.9	57.5	7.6	11.7
$Gebz_4$	0.019	-0.030	195.	60.8	4.3	65.1	96.5	56.9	8.2	12.6
SnM_4	0.131	-0.053	218.	49.7	14.0	63.7	-4.6	50.4	13.3	20.9
SnE_4	0.136	-0.047	218.	49.7	14.0	63.7	-10.7	47.0	16.7	26.2
SnP_4	0.139	-0.045	218.	49.7	14.0	63.7	-34.5	48.1	15.6	24.5
SnB_4	0.140	-0.044	218.	49.7	14.0	63.7	-53.0	47.8	15.9	25.0
PbM_4	0.134	-0.053	230.	39.7	13.6	53.3	32.6	34.7	18.6	34.9
PbE_4	0.139	-0.048	230.	39.7	13.6	53.3	26.3	31.4	21.9	41.1
PM_3	0.042	-0.048	184.1	65.0	8.1	73.1	-22.5	63.9	9.2	12.6
PE_3	0.046	-0.045	184.1	64.9	8.3	73.2	-11.8	55.4	17.8	24.3
AsM_3	-0.065	-0.037	196.	57.6	2.2	59.8	2.8	54.4	5.4	9.1
AsE_3	-0.066	-0.040	196.	57.6	2.2	59.8	13.4	45.9	13.9	23.2
SbM_3	0.064	-0.050	(220.)	48.4	8.6	57.0	7.7	49.7	7.3	12.8
SbE_3	0.069	-0.046	(220.)	48.4	8.8	57.2	11.6	43.5	13.7	24.0
BiM_3	0.110	-0.054	(220.)	48.3	12.4	60.7	46.1	32.5	28.2	46.4
BiE_3	0.116	-0.048	(220.)	48.3	12.4	60.7	51.6	25.8	34.9	57.5

Table 11:2

Bond Energy Data for Phenyl Compounds of Some Elements

cmpd	charge E	C	R_0	E_c	E_i	Ecalc	$\Delta Hf°$	C-E	Wkng	%W
HgPh₂	0.196	-0.036	(207.)	25.9	18.6	44.5	93.8	27.0	17.6	39.4
BPh₃	0.163	-0.032	156.	73.4	20.9	94.3	31.1	100.7	-6.4	-6.8
GePh₄	0.022	-0.028	195.	60.8	4.3	65.1	104.7	62.7	2.4	3.7
SnPh₄	0.155	-0.031	218.	49.7	14.0	63.8	136.5	50.4	13.4	21.0
PbPh₄	0.158	-0.031	230.	39.7	13.6	53.3	169.6	35.8	17.5	32.9
PPh₃	0.065	-0.027	184.	64.9	8.3	73.2	78.5	65.5	7.7	10.5
AsPh₃	-0.052	-0.026	196.	57.6	2.2	59.8	97.6	58.1	1.7	2.9
SbPh₃	0.085	-0.030	(220.)	48.4	8.8	57.2	104.1	52.9	4.3	7.5
BiPh₃	0.134	-0.032	(220.)	48.2	12.5	60.7	138.6	37.0	23.7	39.0

well known to present special experimental problems, and it is possible that the experimental heat of formation for boron triphenyl, which like those of all the compounds of Table 11:2 is positive, is not large enough.] Although it seems reasonable to choose the M-C bond as alone responsible for this difference, there is no proof that this is so.

The experimental data of Tables 11:1 and 11:2 are probably not sufficiently precise to justify any conclusions concerning relative values for different alkyl groups, or even phenyl. Therefore, wherever possible, data for the alkyl derivatives of a given element were averaged, and the averages used in Figures 11:1 and 11:2 to reveal the general trends. Figure 11:1 shows the bond weakening according to major groups and periods of the periodic table. Figure 11:2 does the same for percentage of weakening, assuming the calculated M-C energy to be the "correct" unweakened value. One may note that in the compounds of "inert-shell" elements, the bond weakening is greater in the alkyls of the less electronegative elements. This is also true in the compounds of the "18-shell" elements, except for those beyond the filling of the 4f orbitals. This corresponds to more polar bonds to

Figure 11:1

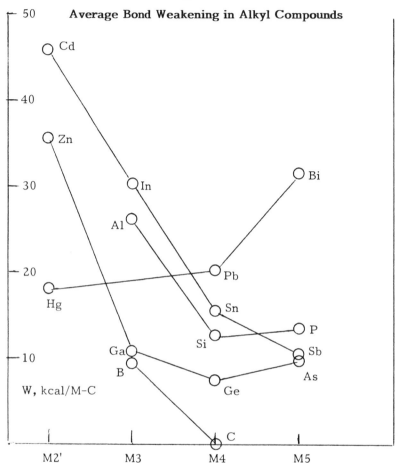

carbon, within any particular group, which suggests that weakening is greater where the carbon is more negative. A certain degree of similarity to chemistry of negative hydrogen is noted. The theory of polar covalence applied to compounds of negative hydrogen gives calculated bond energies higher than those experimentally determined, although bond energies in compounds of positive hydrogen can be accurately calculated.

It is therefore of interest to consider the effect of

Figure 11:2

Percentage of Bond Weakening in Alkyl Compounds

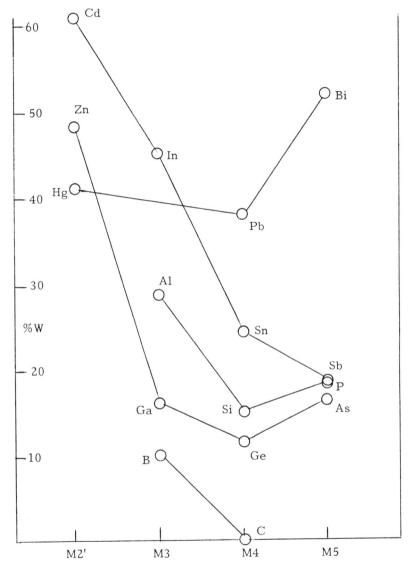

Figure 11:3

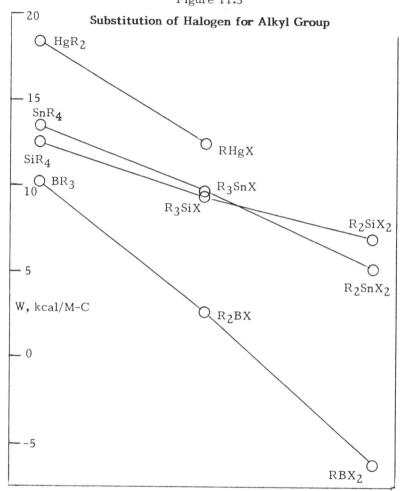

substituting halogen for alkyl group on the remaining M-C bonds. This should decrease the negative charge on carbon, and therefore possibly diminish the weakening. The data for such alkyl metal halides are very limited, but such as they are, bear out this reasoning. This may be seen in Table 11:3 and graphically in Figure 11:3. The expected reduction

Table 11:3

Bond Weakening in Alkyls and Derivatives

cmpd	charge M	charge C	av BE(M–C)	W/bond		%W	
HgM$_2$	0.164	−0.065	44.5	16.9	} 18.3	38.1	} 41.2
HgE$_2$	0.176	−0.055	44.5	19.7		44.3	
MHgCl	0.206	−0.028	44.5	10.9		24.4	
MHgBr	0.191	−0.041	44.5	12.5		28.1	
MHgI	0.163	−0.066	44.5	10.5		23.7	
EHgCl	0.201	−0.032	44.5	12.5	} 12.4	28.0	} 23.7
EHgBr	0.191	−0.041	44.5	14.6		32.7	
EHgI	0.173	−0.057	44.5	13.5		30.5	
SiM$_4$	0.200	−0.057	83.5	11.0	} 12.6	13.2	} 15.1
SiE$_2$	0.208	−0.050	83.5	14.1		16.9	
M$_3$SiCl	0.220	−0.040	83.5	9.0	} 9.5	10.8	} 11.4
M$_3$SiBr	0.214	−0.045	83.5	9.9		11.9	
MSiHCl$_2$	0.263	−0.002	83.5	7.6	} 7.0	9.1	} 8.3
M$_2$SiCl$_2$	0.251	−0.012	83.5	6.3		7.5	
M$_4$Sn	0.131	−0.053	63.8	13.3		20.9	
M$_3$SnBr	0.145	−0.040	63.8	9.6	} 9.7	15.0	} 15.1
M$_3$SnI	0.133	−0.050	63.8	9.7		15.2	
M$_2$SnCl$_2$	0.183	−0.005	63.8	5.2		8.1	
BM$_3$	0.138	−0.049	94.5	6.8	} 9.9	7.2	} 10.3
BE$_3$	0.145	−0.049	94.5	9.3		9.2	
BB$_3$	0.149	−0.045	94.5	13.7		14.6	
B$_2$BCl	0.158	−0.037	94.5	1.3	} 2.4	1.4	} 2.5
B$_2$BI	0.149	−0.045	94.5	2.0		2.1	
B$_2$BBr	0.155	−0.040	94.5	3.8		4.0	
MBF$_2$	0.266	0.061	94.5	−2.7	} −6.2	−2.8	} −6.5
EBF$_2$	0.231	0.030	94.5	−9.6		−10.1	

in weakening even occurs in boron alkyls, but here the sub-stitution of two halogen atoms for alkyl groups produces an apparent strengthening of the remaining B-C bond. This seems very unlikely, suggesting perhaps a fundamental in-accuracy in all the data of these boron compounds.

Although it is tempting to try to find a quantitative explanation of the M-C bond weakening in terms of partial charges on M and C, there are three examples in Table 11:3 that suggest caution. These are the iodides, CH_3HgI, C_2H_5HgI, and $(CH_3)_3SnI$. In these, the partial charges on M and C are almost identical with those on the unsubstituted alkyls, yet the weakening is reduced. These suggest that M-C bond weak-ening is reduced as the number of M-C bonds is reduced, independent of the effect on partial charges. No explanation is offered.

Unfortunately, it appears that further study must await more accurate data, for many more compounds. This chapter only suggests the nature of the problem as describable from evidence available at present, which is insufficient.

Chapter 12

BOND DISSOCIATION AND THE
REORGANIZATIONAL ENERGIES OF FREE RADICALS

The contributing bond energies (CBE's) that are so important in determining whether atoms will rearrange do not always provide adequate information about the strengths of individual bonds. To measure bond strength, the bond dissociation energy (BDE) is determined. This is the same as the contributing bond energy only for diatomic molecules. Any molecule containing more than one bond will form two free radicals when one of the bonds is broken, at least one of which still contains one or more chemical bonds. It is important to recognize that the liberation of a formerly bonding electron to a polyatomic radical must involve a reorganization of the electronic system of that radical, to adjust for the fact that that electron is no longer tied up in a bond. If possible, the liberated bonding electron will be employed in strengthening the remaining bonds in the radical. If so, energy released by this reorganization will make the original bond breaking easier, requiring that much less outside energy. On the other hand, if the released bonding electron cannot be accommodated within the residual bonding, it is likely to interfere, weakening that bonding. This has the result that breaking the original bond requires more energy than expected from the contributing bond energy value, enough extra to weaken the residual bonds.

The relationship between BDE and CBE is therefore the following:

$$BDE = CBE + E_R(I) + E_R(II) \qquad (12\text{-}1)$$

$E_R(I)$ and $E_R(II)$ are the reorganizational energies of the radicals I and II. They are zero if the radical is a free atom but must be evaluated if the radical is polyatomic and therefore contains residual bonds.

For example, although the contributing bond energy in a molecule of water is about 111 kcal (464 kJ) per mole, it costs about 121 kcal (506 kJ) to break one hydrogen atom loose, because the liberated electron on the hydroxyl group weakens the remaining bond by about 10 kcal. There is no way of breaking the first bond without simultaneously weakening the second, so the bond dissociation requires more energy than anticipated from the CBE. The reorganizational energy of the hydroxyl radical is approximately 10 kcal (42 kJ) per mole. For this example it is not easy to explain exactly how the liberated electron weakens the remaining bond. Perhaps it enters the region between oxygen nucleus and bonding electrons, reducing the effective nuclear charge sensed by the bonding electrons. Whether or not the phenomenon can be satisfactorily explained at this time, **it is very real and very significant in determining the bond dissociation energy.**

The simplest kind of reorganizational energy to explain is that involved when the free radical is able to rearrange to form a stable molecule. We have already seen an example of that, where breaking off an oxygen atom from a molecule of CO_2 does not require the full 192 kcal (803 kJ) of the CBE, but only 127 kcal (531 kJ). This is because the liberated electron on the CO radical can enter the bonding system to convert the residual double bond to a triple bond of energy 257 kcal (1075 kJ) per mole, in the very stable carbon monoxide molecule, from its original value of 192. The 65 kcal (272 kJ) so released reduces by that amount the energy required for dissociation of the first oxygen atom, and is, in fact, the reorganizational energy of the carbonyl diradical, $=C=O$.

Since many chemical reactions appear to proceed through free radical mechanisms, it is especially important to be able to evaluate these reorganizational energies, as a means of understanding and predicting the bond dissociation energies. There are two principal ways of calculating bond dissociation energies. One is to subtract the standard heat of formation of a compound from the sum of the standard heats of formation of the two radicals:

$$BDE = \Delta Hf^\circ(I) + \Delta Hf^\circ(II) - \Delta Hf^\circ(I+II) \qquad (12\text{-}2)$$

The other is to add the reorganizational energies, E_R, of the radicals I and II to the contributing bond energy, as shown by equation (12-1). These two equations were used in evaluation of the reorganizational energies in the following way. Bond dissociation energies were calculated from the heats of formation, using values for radicals tabulated by Kerr and Trotman-Dickenson (A15), supplemented by more recent data (A16-A20) and the heats of formation of gaseous molecules recommended by Cox and Pilcher (A13). CBE's were obtained from the data of Tables 6:2, 10:5, and 10:6, and used in equation (12-2) to determine the reorganizational energies. First, compounds of hydrocarbon radicals with hydrogen and halogens were used to evaluate the reorganizational energies of these radicals approximately, since the E_R values for H, F, Cl, Br, and I are of course zero. These values were then refined by application to the bond dissociation energies of hydrocarbons composed of two such radicals. This involved all possible combinations of the following radicals: methyl, ethyl, propyl, isopropyl, t-butyl, vinyl, allyl, phenyl, and benzyl. The reorganizational energies thus obtained were an average of twelve different determinations. These hydrocarbon values were then employed in evaluating the reorganizational energies of functional group radicals or radicals containing such groups.

The radical reorganizational energies thus evaluated were applied to calculating 268 different bond dissociation energies. The average difference between bond dissociation energies obtained in this way and those from heats of formation was only 0.5 kcal per mole. The heats of formation and the reorganizational energies of 32 radicals are given in Table 12:1. A sampling of their application in the estimation of bond dissociation energies is presented in Table 12:2. For reasons not yet clear, there are, among the 268 bonds studied, a total of 30 for which the difference between BDE's calculated by the two methods differ by somewhat more than 1 kcal per mole. These show no consistency, 14 values being higher and 16 lower.

It will be recognized that the accuracy of this work is completely dependent on the validity of the experimental heats of formation and on the correct evaluation of the contributing bond energies. Overall, the results are regarded as very satisfactory. A variation of the method of evaluating radical reorganizational energy is to subtract the calculated contributing bond energies in the radical, where these can be determined, from the same energies in the molecule. Results are in good agreement with those obtained by the averaging process but are only possible for a limited number

Table 12:1

Heats of Formation and Reorganizational Energies of Free Radicals

radical	$\Delta Hf°$	E_R	radical	$\Delta Hf°$	E_R
H	52.1	0.	-CHO	7.7	-6.0
F	18.9	0.	CH_3CO-	-5.8	-6.7
Cl	29.0	0.	C_6H_5CO-	26.1	-5.6
Br	26.7	0.	-COOH	-53.3	-5.0
I	25.5	0.	CH_3COO-	-49.6	1.3
CH_3-	35.1	3.9	NH_2-	47.2	17.8
C_2H_5-	28.2	2.0	CH_3NH-	45.4	11.9
nC_3H_7-	22.6	1.5	$-NO_2$	6.0	-10.3
iC_3H_7-	20.6	0.9	-OH	9.9	9.8
tC_4H_9-	10.5	-1.7	CH_3O-	3.8	-4.4
$CH_2=CH-$	68.	10.0	C_2H_5O-	-4.1	-4.3
$CH_2=CH-CH_2-$	41.4	-10.1	C_6H_5O-	11.4	-24.2
C_6H_5-	77.7	11.2	-SH	33.1	2.9
$C_6H_5CH_2-$	45.1	-13.9	CH_3S-	34.2	5.1
$HOCH_2-$	-6.2	-1.6	C_6H_5S-	56.8	-4.4
-CN	101.	29.0	CH_3SO_2-	-57.2	-4.7

of radicals and are not regarded as reliable as those that can be based on a number of examples instead of just one. Table 12:3 gives bond dissociation data for methyl compounds, to illustrate the general procedure.

Ideally we should be able to examine the composition and structure of any radical and from these estimate the reorganizational energy. In some instances we can, for if a radical can rearrange to form a stable molecule, the necessary data are usually available and the reorganization can be understood. Interpretation is not so easy for other examples, where it might be difficult even to predict whether the reorganization should be exothermic or endothermic. The stabilizing reorganization of the allyl radical is of course associated with the electron delocalization, which makes

Table 12:2

Representative Bond Dissociation Energies

bond	dissociation energy	
	calc	exp
CH_3Cl	84.8	84.7
C_2H_5-Br	70.4	70.2
$CH_3-CH(CH_3)_2$	88.3	88.1
$C_2H_5-CHCH_2$	96.4	96.4
$C_6H_5-CH(CH_3)_2$	97.5	97.3
$C_2H_5-CH_2OH$	83.2	83.6
nC_3H_7-CN	114.3	113.3
CH_2CHCH_2-CHO	68.3	68.5
$CH_3CO-C_6H_5$	92.5	92.7
$C_6H_5CO-tC_4H_9$	75.4	75.3
$C_6H_5CH_2-COOH$	67.2	66.7
$CH_3COO-CHCH_2$	96.7	96.7
$iC_3H_7-NH_2$	87.8	87.8
CH_3NH-CH_3	84.8	85.2
$C_6H_5-NO_2$	69.1	69.7
C_2H_5-OH	93.8	94.3
$CH_3O-CH_2C_6H_5$	67.8	68.4
$C_6H_5O-CH_2CHCH_2$	51.8	52.2
$CH_3S-nC_3H_7$	76.2	76.3

the two C-C bonds equal to one another and intermediate between single and double. The NO_2 radical is stabilized by changing to the NO_2 molecule. Release of an electron to an atom attached to a phenyl group appears to strengthen the bonding in the benzyl, phenoxy, and benzoyl radicals and even the C_6H_5S radical. No doubt this is related to favorable interaction involving the pi electrons of the ring carbon atoms. The data of Table 12:1 inspire many questions, to which some day satisfactory answers will probably be found. For example,

Table 12:3

Representative Dissociations of Methyl Radicals

II	$\Delta Hf°(I)(CH_3)$	$\Delta Hf°(II)$	$\Delta Hf°(cmpd)$	BDE exp
	CBE	$E_R(CH_3)$	$E_R(II)$	BDE calc
H	35.1	52.1	-17.9	105.1
	99.4	3.9	0.0	103.3
Cl	35.1	29.0	-20.6	84.7
	80.9	3.9	0.0	84.8
I	35.1	25.5	3.3	57.3
	53.4	3.9	0.0	57.3
C_2H_5	35.1	28.2	-24.8	88.1
	82.8	3.9	2.0	88.7
iC_3H_7	35.1	20.6	-32.4	88.1
	83.5	3.9	0.9	88.3
$CH_2=CH$	35.1	68.	4.9	98.2
	84.4	3.9	10.0	98.3
C_6H_5	35.1	77.7	12.0	100.8
	85.8	3.9	11.2	100.9
CHO	35.1	7.7	-39.7	82.5
	84.4	3.9	-6.0	82.3
C_6H_5CO	35.1	26.1	-20.7	81.9
	83.8	3.9	-5.6	82.1
CH_3COO	35.1	-49.6	-97.9	83.4
	78.3	3.9	1.3	83.5
NH_2	35.1	47.2	-5.5	87.8
	65.7	3.9	17.8	87.4
SH	35.1	33.1	-5.5	73.7
	66.7	3.9	2.9	73.5

Table 12:4

Comparison of Phenyl and Benzyl Bond Dissociation Energies

other radical	phenyl BDE	benzyl BDE	difference
H	110.0	84.9	25.1
F	124.4	93.7	30.7
Cl	94.5	67.7	26.8
Br	79.3	54.7	24.6
I	64.2	40.1	24.1
CH_3	100.9	72.9	28.0
C_2H_5	98.9	71.2	27.7
C_3H_7	98.5	70.7	27.8
iC_3H_7	97.4	69.9	27.5
tC_4H_9	93.7	67.2	26.5
$CH_2=CH$	108.4	80.7	27.7
$CH_2=CH-CH_2$	86.9	58.8	28.1
C_6H_5	109.9	82.9	27.0
$C_6H_5CH_2$	82.9	58.9	24.0
$HOCH_2$	95.5	67.2	28.3
CN	128.0	99.9	28.1
CHO	93.8	64.6	29.2
CH_3CO	92.6	63.0	29.6
C_6H_5CO	93.7	65.1	28.6
COOH	93.7	67.0	26.7
CH_3COO	99.3	69.2	30.1
NH_2	104.0	71.6	32.4
CH_3NH	98.2	68.3	29.9
NO_2	69.4	43.2	26.2
OH	110.8	78.5	32.3
CH_3O	99.1	68.1	31.0
C_2H_5O	98.9	68.3	30.6
C_6H_5O	79.9	48.2	31.7

Table 12:4 cont.

other radical	phenyl BDE	benzyl BDE	difference
SH	84.0	56.9	27.1
CH_3S	88.2	60.5	27.7
C_6H_5S	78.9	51.5	27.9
CH_3SO_2	80.6	52.6	28.0
E_R	11.2	-13.9	25.1

why does release of one bonding electron to the phenyl group weaken its bonding?

The very high positive reorganizational energy of the cyanide radical suggests that its liberation is accompanied by a reduction in bond multiplicity, similar to that of removing two hydrogen atoms from acetylene, when the triple bond changes to double in the C_2 radical. This is supported by the bond energy of only about 184 kcal (770 kJ) in CN radical, compared to 210.7 from Table 10:5. The difference, 26.7 kcal, is close to the average E_R value of 29 kcal (121 kJ) given in Table 12:1.

Fortunately, without the perfection of understanding, we still can make valuable practical use of many chemical observations, and the reorganizational energies of free radicals are no exception. Current textbooks of organic chemistry commonly include a list of common bond dissociation energies. A list of reorganizational energies of free radicals would be far more informative and useful, because of their general applicability. For example, every bond to a phenyl group will require 11.2 kcal (46.9 kJ) more to break it than would be expected from the CBE, no matter what the bond is. This helps to explain the relative unreactivity of functional groups attached to the benzene ring, compared with aliphatic derivatives. The difference between bond strengths in phenyl and benzyl compounds is similarly always in the neighborhood of 25 kcal per mole (105 kJ), since the reorganizational energies are +11 for phenyl and -14 for benzyl. This is illustrated by the data of Table 12:4, wherein bond dissociation energies are compared for a number of phenyl and benzyl derivatives.

It will be observed that the differences are somewhat variable and usually greater than 25 kcal (105 kJ), averaging about 28 kcal (117 kJ). This is because as a rule the contributing bond energies of bonds of functional groups to phenyl are somewhat larger, but to a varying degree, than those of bonds to benzyl.

The sequence of decreasing positive and increasing negative reorganizational energies in hydrocarbon radicals is essentially the same as that long recognized: phenyl, vinyl, methyl, ethyl, propyl, isopropyl, t-butyl, allyl, and benzyl. This is the order of increasing ease of formation of the free radicals. It can be very useful in interpreting the nature of reactions of hydrocarbons by free radical mechanisms.

As soon as the influence of unit charge on bond energy in positive and negative ions becomes understood and can be quantitatively evaluated, a similar study of the reorganization of polyatomic ions can be made, which should prove very useful also.

Chapter 13

BOND DISSOCIATION IN INORGANIC MOLECULES

Bond dissociation in organic molecules was discussed in Chapter 12. Therein were included a number of radicals which occur also in a number of inorganic compounds, such as OH, NH_2, and NO_2. In this chapter a number of additional radicals will be studied along with the inorganic applications of those previously evaluated.

HALIDES AND SUBHALIDES

The largest source of information is heats of formation and bond energies in a number of inorganic halides and sub-halides. Although these data, especially for the subhalides which commonly are known only in the gaseous state, are not very precise, a helpful observation has been made. When the bond energies of the normal halides are plotted against those of the subhalides in the same major group, in a particular oxidation state, including the complete estimated range of possible experimental error, they appear to be linear, within the likely limits of error. This is true for each particular group of the periodic table, and seems independent of the identity of the particular element in that group or the halogen.

For example, if the average contributing bond energy of the (IV) fluorides, chlorides, bromides, and iodides is plotted against the average contributing bond energy in the corresponding (II) halides, for a particular major group, a linear relationship is found from which a more precise value for the subhalide can be estimated. The difference between the values for the two oxidation states represents the reorganizational energy for the breaking away of one halogen atom, or in this example the average reorganizational energy for the breaking away of two halogen atoms. The parameters for the appropriate linear equations are summarized in Table 13:1.

Both the experimental and calculated contributing bond energies, based on the theory of polar covalence, are included in this study, and the equations of Table 13:1 are used to calculate from these two values the corresponding values for the lower oxidation state. Thus are obtained two

Table 13:1

Relationship of Bond Energies in Normal and Subhalides

Group		a) Substate BE = A x Normal BE + B				
		b) Sum of E_R's = A' x Normal Be + B				
M2	a)	I	1.16	II	-28.4
	b)	I	-0.16	II	28.4
M2'	a)	I	1.12	II	-30.3
	b)	I	-0.12	II	30.3
M3	a)	I	0.84	III	38.7
	b)	I	0.16	III	-38.7
M4	a)	I	1.03	IV	3.4
	b)	I	-0.03	IV	-3.4
	a)	I	1.11	II	-15.5
	b)	I	-0.11	II	15.5
	a)	II	0.93	IV	16.8
	b)	II	0.07	IV	-16.8
M5	a)	I	0.85	III	15.6
	b)	I	0.15	III	-15.6
T4	a)	I	0.86	IV	19.2
	b)	I	0.14	IV	-19.2
	a)	II	0.91	IV	23.5
	b)	II	0.09	IV	-23.5
	a)	III	1.09	IV	-1.1
	b)	III	-0.09	IV	1.1

reorganizational energies, from which, together with energies for the normal halide, an average bond dissociation energy is determined. As will be discussed presently, these BDE values are usefully informative, and in some applications shed light on interesting and important chemical properties of halides. Table 13:2 gives BDE's for 120 halides.

The basic data for estimating the reorganizational energies of major group subhalide molecules are given in

Table 13:2

Bond Dissociation Energies of Inorganic Halides

(for one bond or average of 2 or 3 bonds)

cmpds	BDE	cmpds	BDE	cmpds	BDE
BeF_2-F	156	$TlCl_3$-Cl	30	$SnCl_2$-Cl	92
$BeCl_2$-Cl	121	$TlBr_3$-Be	17	$SnBr_2$-Br	83
$BeBr_2$-Br	107	TlI_3-I	1	SnI_2-I	72
BeI_2-I	90	CF_4-F_2	109	PbF_2-F	94
MgF_2-F	138	CCl_4-Cl_2	67	$PbCl_2$-Cl	81
$MgCl_2$-Cl	115	SiF_4-F_2	134	$PbBr_2$-Br	72
$MgBr_2$-Br	107	$SiCl_4$-Cl_2	84	PbI_2-I	62
MgI_2-I	89	$SiBr_4$-Br_2	69	NF_3-F	63
CaF_2-F	141	SiI_4-I_2	47	PF_3-F	124
$CaCl_2$	116	GeF_4-F_2	103	PCl_3-Cl	73
$CaBr_2$-Br	107	$GeCl_4$-Cl_2	66	PBr_3-Br	58
CaI_2-I	93	$GeBr_4$-Br_2	53	PI_3-I	38
SrF_2-F	140	GeI_4-I_2	37	AsF_3-F	(100)
$SrCl_2$-Cl	116	SnF_4-F_2	98	$AsCl_3$-Cl	(61)
$SrBr_2$-Br	107	$SnCl_4$-Cl_2	63	$AsBr_3$-Br	(49)
SrI_2-I	93	$SnBr_4$-Br_2	51	SbF_3-F	102
BaF_2-F	142	SnI_4-I_2	37	$SbCl_3$-Cl	69
$BaCl_2$-Cl	(118)	PbF_4-F_2	65	$SbBr_3$-Br	55
$BaBr_2$-Br	102	$PbCl_4$-Cl_2	48	SbI_3-I	37
BaI_2-I	(98)	$PbBr_4$-Br_2	37	$BiCl_3$-Cl	61
ZnF_2-F	111	PbI_4-I_2	24	$BiBr_3$-Br	50
$ZnCl_2$-Cl	100	CF_4-F	110	BiI_3-I	35
$ZnBr_2$-Br	88	CCl_4-Cl	72	TiF_4-F_3	128
ZnI_2-I	74	SiF_4-F	138	$TiCl_4$-Cl_3	95
CdF_2-F	110	$SiCl_4$-Cl	88	$TiBr_4$-Br_3	81
$CdCl_2$-Cl	100	$SiBr_4$-Br	74	TiI_4-I_3	65
$CdBr_2$-Br	89	SiI_4-I	55	ZrF_4-F_3	142
CdI_2-I	77	GeF_4-F	104	$ZrCl_4$-Cl_3	108
HgF_2-F	93	$GeCl_4$-Cl	72	$ZrBr_4$-Br_3	94
$HgCl_2$-Cl	78	$GeBr_4$-Br	60	ZrI_4-I_3	77
$HgBr_2$-Br	70	GeI_4-I	45	HfF_4-F_3	143
HgI_2-I	60	SnF_4-F	101	$HfCl_4$-Cl_3	109
BF_3-F	143	$SnCl_4$-Cl	68	TiF_4-F_2	129
BCl_3-Cl	84	$SnBr_4$-Br	58	$TiCl_4$-Cl_2	88
BBr_3-Br	66	SnI_4-I	45	$TiBr_4$-Br_2	72
BI_3-I	39	PbF_4-F	71	TiI_4-I_2	53
AlF_3-F	127	$PbCl_4$-Cl	55	ZrF_4-F_2	146
$AlCl_3$-Cl	79	$PbBr_4$-Br	46	$ZrCl_4$-Cl_2	105
$AlBr_3$-Br	61	PbI_4-I	34	$ZrBr_4$-Br_2	87
AlI_3-I	41	CF_2-F	128	ZrI_4-I_2	68
GaF_3-F	99	SiF_2-F	147	HfF_4-F_2	146

Table 13:2 cont.

cmpds	BDE	cmpds	BDE	cmpds	BDE
$GaCl_3$-Cl	(53(	$SiCl_2$-Cl	109	$HfCl_4$-Cl_2	106
$GaBr_3$-Br	(41)	$SiBr_2$-Br	97	TiF_4-F	140
GaI_3-I	20	SiI_2-I	80	$TiCl_4$-Cl	98
InF_3-F	72	GeF_2-F	123	$TiBr_4$-Br	81
$InCl_3$-Cl	46	$GeCl_2$-Cl	96	TiI_4-I	61
$InBr_3$-Br	35	$GeBr_2$-Br	85	ZrF_4-F	158
InI_3-I	19	GeI_2-I	72	$ZrCl_4$-Cl	115
TlF_3-F	54	SnF_2-F	119	$ZrBr_4$-Br	97

Tables 13:3-9. The following general observations appear to be true: (1) Whether negative or positive, the reorganizational energies increase from fluoride to iodide for any given element. (2) When halogen atoms are split off from halide molecules, the remaining bonds are **strengthened but lengthened** if the splitting liberates an even number of outermost electrons, and **weakened but shortened** if the number of liberated electrons is odd. Data are insufficient to establish these observations for all possibilities, but available data are uniformly consistent with them.

For example, when $HgCl_2$ becomes HgCl, loss of the chlorine atom corresponds to the release of one electron. The contributing bond energy decreases from about 54 kcal (226 kJ) per mole of bonds in $HgCl_2$ to 30 kcal (126 kJ) per mole in HgCl. The bond length, however, changes from 231 pm in $HgCl_2$ to 223 in HgCl. In contrast, loss of two chlorine atoms from a molecule of $AlCl_3$ releases two formerly bonding electrons. The bond energy increases from about 102 kcal (428 kJ) in $AlCl_3$ to 124 kcal (519 kJ) in AlCl, but the bond length increases from 206 to 213 pm.

Since bond energy is essentially the result of electrostatic attractions which increase with decreasing distance, any **increase** in bond energy would be expected to correspond to a **shortening** of the bond, and in normal compounds this is true. The reverse trend observed in the subhalides suggests something unusual or at least unexpected. Perhaps when an atom is exerting less than its normal valence, it may no longer be regarded as spherical, thus affecting the bond length in a manner opposite to that anticipated. It would of course

Table 13:3

Bond Dissociation and Reorganizational Energies of M2 Halides

cmpd	R_o	CBE exp calc	cmpd	R_o	BE calc	exp	E_R
BeF_2	140	152.5 150.5	BeF	136	148.5 146.2	128-148	4.0 4.3
$BeCl_2$	177	110.6 108.8	BeCl	178	99.9 97.8	90-95	10.7 11.0
$BeBr_2$	191	92.9 93.3	BeBr	(193)	79.4 79.8	71-111	13.5 13.5
BeI_2	212	71.9 74.0	BeI	213	55.0 57.4		16.9 16.6
MgF_2	177	131.3 130.6	MgF	175	123.9 123.1	117-120	7.4 7.5
$MgCl_2$	218	103.4 102.3	MgCl	220	91.5 90.3	81-97	11.9 12.0
$MgBr_2$	234	90.3 89.0	MgBr	236	76.3 74.8	64-94	14.0 14.2
MgI_2	252	71.2 71.9	MgI	(256)	54.2 55.0		17.0 16.9
CaF_2	210	134.0 133.7	CaF	202	127.0 126.7	121-131	7.0 7.0
$CaCl_2$	251	106.7 102.1	CaCl	244	95.4 90.0	92-98	11.3 12.1
$CaBr_2$	267	95. 92.2	CaBr	(260)	81.8 78.6	71-82	13.2 13.6
CaI_2	288	78.3 75.9	CaI	(280)	62.4 59.6	53-83	15.9 16.3
SrF_2	220	129.8 135.9	SrF	(212)	122.2 129.2	128-131	7.6 6.7
$SrCl_2$	267	106.6 102.8	SrCl	(258)	95.3 90.8	94-100	11.3 12.0
$SrBr_2$	282	95.3 92.6	SrBr	(274)	82.1 79.0	74-84	13.2 13.6
SrI_2	303	77.6 76.6	SrI	(294)	61.6 60.5	53-73	16.0 16.1
BaF_2	232	137.9 131.7	BaF	(224)	131.6 124.4	138-142	6.3 7.3

Table 13:3 cont.

cmpd	R_o	CBE exp calc	cmpd	R_o	BE calc	exp	E_R
$BaCl_2$	282	113.1 99.7	BaCl	(275)	102.8 87.3	103–109	10.3 12.4
$BaBr_2$	299	98.8 88.6	BaBr	(291)	86.2 74.4	86–90	12.6 14.2
BaI_2	320	85.8 79.1	BaI	(311)	71.1 63.4		14.7 15.7

be very desirable to understand the relationship between the number of electrons released and the strengthening or weakening of the bond, as well as the effect on bond length.

Of particular interest is the change from the (IV) to the (II) state in elements known for their tendency to leave two outer electrons paired, especially tin and lead. In the past it was assumed that the homonuclear energy and covalent radius of the metal remain unchanged, and by back calculation from experimental bond energies it was possible to evaluate the (II) electronegativities. These were found to be lower than for the (IV) state, making bonds to more electronegative elements more polar and hence stronger. However, the assumptions now seem dubious, weakening confidence in the (II) electronegativity values. The bonds **are** stronger in the (II) state, as seen in Table 13:7, despite longer bonds in the (II) state. However, from Si through Pb the increase from (IV) to (II) is not very different, in the range of 10–14 kcal (48–59 kJ) per mole of bonds. There is no obvious clue to the increasing stability of the (II) state down the group.

Thermochemical data do show, however, how the relative stability of the (II) state increases from Sn to Pb. The atomization energy of $SnCl_4$(liq) is 310.4 kcal (1299 kJ) per mole, whereas the sum for $SnCl_2$(c), 207.9 kcal (870 kJ) and Cl_2(g) 58.0 kcal (242.6 kJ) is only 265.9 kcal (1112.5 kJ). The $SnCl_4$ is 44.5 kcal (186 kJ) per mole **more** stable than its equivalent as $SnCl_2$(c) and Cl_2(g). In contrast, the atomization energy of $PbCl_4$(liq) is 241.3 kcal (1009.6 kJ), but the sum for $PbCl_2$(c) + Cl_2(g) is 190.5 + 58.0 or 247.5 kcal (1035.5 kJ). Here $PbCl_4$(liq) is **less** stable than $PbCl_2$(c) + Cl_2(g) by 6.2 kcal (25.9 kJ).

Table 13:4

Bond Dissociation and Reorganizational Energies of M2' Halides

cmpd	R_o	CBE exp calc	cmpd	R_o	BE calc	exp	E_R
ZnF_2	181	92.0 90.5	ZnF	174	72.7 71.1	73–103	19.3 19.4
$ZnCl_2$	205	78.7 78.5	ZnCl	199	57.8 57.6	50–59	20.9 20.9
$ZnBr_2$	221	66.0 66.0	ZnBr	(215)	43.6 43.6	27–41	22.4 22.4
ZnI_2	238	50.1 51.3	ZnI	(231)	25.7 28.8	26–40	24.3 22.5
CdF_2	197	76.8? 90.6	CdF		55.7 71.2	68–78	21.1 19.4
$CdCl_2$	221	68.4? 78.6	CdCl		46.3 57.7	48–50	22.1 20.9
$CdBr_2$	237	58.5? 66.9	CdBr		35.2 44.6	15–61	23.3 22.3
CdI_2	255	45.6? 52.5	CdI		20.8 28.5	28–38	24.8 24.0
HgF_2	196	61.4? 70.9	HgF	189	38.5 49.1	22–40	22.9 21.8
$HgCl_2$	231	53.9 53.7	HgCl	223	30.1 29.8	22–26	23.8 23.9
$HgBr_2$	244	44.3 45.0	HgBr		19.3 20.1	16–18	25.0 25.0
HgI_2	261	34.6 32.3	HgI	249	8.5 5.9	9	26.1 26.4

TRANSITIONAL METAL HALIDES

The effects of underlying d electrons and changes in oxidation state on the electronegativities of the transitional elements have not yet been evaluated. Furthermore these elements form relatively few gaseous molecules and there is as yet no way to evaluate their homonuclear single covalent bond energies. For these reasons and others, e.g. lack of

Table 13:5

**Bond Dissociation and Reorganizational Energies of
M3 Halides**

cmpd	R_o	CBE exp calc	cmpd	R_o	BE calc	exp	E_R
BF_3	131	155.7 158.2	BF	126	169.5 171.6	180-186	-13.8 -13.4
BCl_3	175	107.4 104.3	BCl	172	128.9 126.3	126-136	-21.5 -22.0
BBr_3	187	89.3 90.7	BBr	189	113.7 114.9	96-106	-24.4 -24.2
BI_3	203	66.2 67.0	BI		94.3 95.0	86-96	-28.1 -28.0
AlF_3	163	141.5 144.9	AlF	165	157.6 160.4	157-160	-16.1 -15.5
$AlCl_3$	206	101.8 101.7	AlCl	213	124.2 124.1	115-121	-22.4 -22.4
$AlBr_3$	227	85.7 86.1	AlBr	229	110.7 111.0	104-108	-25.0 -24.9
AlI_3	244	68.1 70.0	AlI	254	95.9 97.5	87-89	-27.8 -27.5
GaF_3	188	101.8	GaF	178	124.2	134-142	-22.4
$GaCl_3$	209	86.8? 78.7	GaCl	220	111.6 104.8	112-118	-24.8 -26.1
$GaBr_3$	228	71.9? 66.0	GaBr	235	99.1 94.1	102-110	-27.2 -28.1
GaI_3	244	50.9	GaI	257	81.5	81	-30.6
InF_3	(216)	95.0	InF	199	118.5	117-125	-23.5
$InCl_3$	246	73.8 72.1	InCl	240	100.7 99.3	103-107	-26.9 -27.2
$InBr_3$	258	64.2 63.0	InBr	254	92.6 91.6	95-105	-28.4 -28.6
InI_3	280	50.2 48.8	InI	275	80.8 79.7	80	-30.6 -30.9

Table 13:6

Bond Dissociation and Reorganizational Energies of M4 Halides (IV–I)

cmpd	R_O	CBE exp calc	cmpd	R_O	BE calc	exp	E_R
CF_4	132	117.0 117.5	CF	127	123.9 124.4	123–133	-6.9 -6.9
CCl_4	176	78.0 85.6	CCl	164	83.7 91.6	88–102	-5.7 -6.0
SiF_4	154	142.4 139.5	SiF	160	150.1 147.1	126–132	-7.7 -7.6
$SiCl_4$	202	95.3 93.8	SiCl	206	101.6 100.0	99–101	-6.3 -6.2
$SiBr_4$	215	78.8 81.2	SiBr		84.6 87.0	70–94	-5.8 -5.8
SiI_4	242	59.1 60.4	SiI		64.3 65.6	61–101	-5.2 -5.2
GeF_4	168	112.5 110.0	GeF		119.3 116.7	111–121	-6.8 -6.7
$GeCl_4$	209	78.2 76.6	GeCl		83.9 82.3	77–87	-5.7 -5.7
$GeBr_4$	230	67.1 63.5	GeBr		72.5 68.8	54–68	-5.4 -5.3
GeI_4	249	51.4 48.4	GeI		56.3 53.3		-4.9 -4.9
SnF_4	186	107.4	SnF	194	114.0	109–115	-6.6
$SnCl_4$	231	75.2 73.0	SnCl		80.9 78.6	95–103	-5.7 -5.6
$SnBr_4$	244	63.6 63.7	SnBr		68.9 69.0	80–82	-5.3 -5.3
SnI_4	264	49.8	SnI		54.7	46–66	-4.9
PbF_4	208	77.1 75.5	PbF	201	82.8 81.2	83–87	-5.7 -5.7
$PbCl_4$	243	59.4 61.6	PbCl	220	64.6 66.8	65–79	-5.2 -5.2
$PbBr_4$	258	49.4 51.8	PbBr		54.3 56.8	50–68	-4.9 -5.0

Table 13:7

Bond Dissociation and Reorganizational Energies of M4 Halides (IV–II)

cmpd	R_o	CBE exp calc	cmpd	R_o	BE calc	exp	E_R
CF_4	132	117.0 117.5	CF_2	130	125.6 126.1	124.6	−8.6 −8.6
CCl_4	176	78.0 86.8	CCl_2	176	89.3 97.5	91.5	−11.3 −10.7
SiF_4	154	142.4 139.5	SiF_2	159	149.2 139.5	144.5	−6.8 −7.0
$SiCl_4$	202	95.3 93.8	$SiCl_2$	(205)	105.4 104.0	103.4	−10.1 −10.2
$SiBr_4$	215	78.8 81.2	$SiBr_2$	(217)	90.1 92.3		−11.3 −11.1
SiI_4	242	59.1 60.4	SiI_2	(244)	71.8 73.0		−12.7 −12.6
GeF_4	168	112.5 110.0	GeF_2	(180)	121.4 119.1	124.2	−8.9 −9.1
$GeCl_4$	209	78.2 76.6	$GeCl_2$	(214)	92.3 88.0		−11.1 −11.4
$GeBr_4$	230	67.1 63.5	$GeBr_2$	(237)	79.2 75.9	79.2	−12.1 −12.4
GeI_4	249	51.4 48.4	GeI_2	(255)	64.6 61.8	64.7	−13.3 −13.4
SnF_4	186	107.4	SnF_2	206	116.7	109.	−9.3
$SnCl_4$	231	63.6 73.0	$SnCl_2$	243	86.7 84.7	90.2	−11.5 −11.7
$SnBr_4$	244	63.6 63.7	$SnBr_2$	255	75.9 76.0	77.4	−12.3 −12.3
SnI_4	264	49.8	SnI_2	278	63.1	60.6	−13.3
PbF_4	208	77.1 75.5	PbF_2	213	88.5 87.0	94.3	−11.4 −11.5
$PbCl_4$	243	59.4 61.6	$PbCl_2$	246	72.0 87.0	72.8	−12.6 −12.5
$PbBr_4$	258	49.4 51.8	$PbBr_2$	260	62.7 65.0	62.2	−13.3 −13.2

Table 13:8

Bond Dissociation and Reorganizational Energies of M4 Halides (II–I)

cmpd	BE calc	cmpd	BE calc	E_R
CF_2	125.6, 126.1	CF	123.9, 124.5	1.7
SiF_2	149.2, 146.5	SiF	150.1, 147.1	-0.9,-0.6
$SiCl_2$	105.4, 104.0	SiCl	101.5, 99.9	3.9, 4.1
$SiBr_2$	90.1, 92.3	SiBr	84.6, 87.0	5.5, 5.3
SiI_2	71.8, 73.0	SiI	64.3, 65.5	7.5, 7.5
GeF_2	121.4, 119.1	GeF	119.3, 116.7	2.1, 2.4
$GeCl_2$	92.3, 88.0	GeCl	87.0, 82.2	5.3, 5.8
$GeBr_2$	79.2, 75.9	GeBr	72.5, 68.7	6.7, 7.2
GeI_2	64.6, 61.8	GeI	56.3, 53.1	8.3, 8.7
SnF_2	116.7	SnF	114.0	2.7
$SnCl_2$	86.7, 84.7	SnCl	80.7, 78.5	6.0, 6.2
$SnBr_2$	75.9, 76.0	SnBr	68.9, 68.9	7.0, 7.1
SnI_2	63.1	SnI	54.5	8.6
PbF_2	88.5, 87.0	PbF	82.8, 81.1	5.7, 5.9
$PbCl_2$	72.0, 74.1	PbCl	64.4, 66.8	7.6, 7.3
$PbBr_2$	62.7, 65.0	PbBr	54.3, 56.7	8.4, 8.3
PbI_2	51.5, 53.2	PbI	41.8, 43.6	9.7, 9.6

understanding of bonding in many solids, the theory of polar covalence can be applied at present only tentatively to a few compounds of transitional elements, leaving relatively untouched a very large area of ignorance.

Nevertheless a number of heats of formation are available for different oxidation states, especially for gaseous halides of elements of Group T4 (transitional 4, formerly IVB). It should be interesting to compare these data with those of major group halides. Tables 13:10-13 provide the data and successive bond dissociation energies are summarized

Table 13:9

Bond Dissociation and Reorganizational Energies of M5 Halides (III–I)

cmpd	R_o	CBE exp calc	cmpd	R_o	BE calc	exp	E_R
NF_3	137	66.7	NF	151	72.3	50–72	−5.6
		69.1			74.3		−5.2
PF_3	157	120.3	PF		117.9	82–128	2.4
		123.0			120.2		2.8
PCl_3	204	77.1	PCl	205	81.1	59–79	−4.1
		76.8			80.9		−4.1
PBr_3	220	63.4	PBr	223	69.5	(63)	−6.1
		64.6			70.5		−5.9
PI_3	247		PI				
		46.9			55.5		−8.6
AsF_3	171	105.6	AsF		105.4		0.2
		100.2			100.8		−0.6
$AsCl_3$	216	73.9	AsCl		78.4		−4.5
		66.9			72.5		−5.6
$AsBr_3$	233	67.2	AsBr		67.6		−0.4
		56.0			63.2		−7.2
SbF_3	(190)	103.0	SbF		103.2	82–112	−0.2
		102.0			102.3		−0.3
$SbCl_3$	233	75.1	SbCl		79.4	74–98	−4.4
		71.4			76.3		−4.9
$SbBr_3$	251	63.3	SbBr		69.4	61–89	−6.3
		60.4			66.9		−6.5
SbI_3	272	45.5	SbI		54.3		−8.8
		46.3			55.0		−8.7
$BiCl_3$	248	66.7	BiCl		72.3	68–72	−5.6
		67.1			72.6		−5.5
$BiBr_3$	263	56.0	BiBr		63.2	63.2	−7.2
		57.6			64.6		−7.0
BiI_3	284	43.3	BiI		52.4	52.1	−9.1
		44.2			53.2		−9.0

Table 13:10

**Bond Dissociation and Reorganizational Energies of
T4 Halides (IV–III)**

cmpd	CBE exp	cmpd	CBE calc	CBE exp	E_R
TiF_4	139.7	TiF_3	151.2	151.0	−11.5
$TiCl_4$	102.7	$TiCl_3$	110.8	109.6	−8.1
$TiBr_4$	87.7	$TiBr_3$	94.5	94.0	−6.8
TiI_4	70.2	TiI_3	75.4	74.9	−5.2
ZrF_4	155.3	ZrF_3	168.2	155.7	−12.9
$ZrCl_4$	117.4	$ZrCl_3$	126.9	119.3	−9.5
$ZrBr_4$	101.6	$ZrBr_3$	109.6	109.6	−8.0
ZrI_4	83.5	ZrI_3	89.9	91.7	−6.4
HfF_4	155.7	HfF_3	168.6		−12.9
$HfCl_4$	118.9	$HfCl_3$	128.5		−9.6

in Table 13:14.

It appears that in these compounds, the bond energy increases with every release of an electron from bonding. In other words, each successive reorganizational energy is negative.

OXIDES AS OXIDIZING AGENTS

Bond dissociation and radical reorganizational energies may be used to determine the relative ease with which various oxidizing agents can liberate atomic oxygen.

In ozone, O_3, the average bond energy is 72.3 kcal (302.5 kJ) per mole of bonds, but the breaking away of one oxygen atom permits the two-oxygen residue to reorganize, forming the O_2 molecule with bond energy 119.1 kcal (498.3 kJ). Thus E_R for the O_2 fragment is 72.3 − 119.1 = −46.8 kcal (−195.8 kJ) per mole. Therefore release of one oxygen atom only requires 72.3 − 46.8 = 25.5 kcal (106.7 kJ).

The average bond energy in NO_2 is 112.1 kcal (469.0

Table 13:11

Bond Dissociation and Reorganizational Energies of
T4 Halides (IV–II)

cmpd	CBE exp	cmpd	CBE calc	CBE exp	E_R
TiF_4	139.7	TiF_2	150.6	157.3	-10.9
$TiCl_4$	102.7	$TiCl_2$	117.0	113.5	-14.3
$TiBr_4$	87.7	$TiBr_2$	103.3	104.3	-15.6
TiI_4	70.2	TiI_2	87.4	84.0	-17.2
ZrF_4	155.3	ZrF_2	164.8	158.4	-9.5
$ZrCl_4$	117.4	$ZrCl_2$	130.3	124.0	-12.9
$ZrBr_4$	101.6	$ZrBr_2$	116.0	120.4	-14.4
ZrI_4	83.5	ZrI_2	99.5	106.2	-16.0
HfF_4	155.7	HfF_2	165.2		-9.5
$HfCl_4$	118.9	$HfCl_2$	131.7		-12.8

kJ), but in NO it is 150.7 kcal (630.5 kJ) per mole. Thus removal of one oxygen atom allows the NO fragment to reorganize to nitric oxide molecule, increasing the bond strength by 150.7 - 112.1 = 38.6, so -38.6 kcal (-161.5 kJ) is the E_R for the NO radical. Removal of one oxygen atom from NO_2 thus requires only 112.1 - 38.6 = 73.5 kcal (307.5 kJ) per mole.

As discussed earlier, nitrous oxide seems best represented as having an N"=O" double bond of energy 146.7 kcal (613.8 kJ), and an N'''-N''' energy of 119.4 kcal (499.6 kJ). Once an oxygen atom has been liberated, the N_2 residue can form a true triple bond with energy 225.7 kcal (945.2 kJ). E_R for the N_2 fragment is 119.4 -225.9 = -106.5 kcal (-445.6 kJ) per mole. Releasing the oxygen atom therefore requires only 146.7 - 106.5 = 40.2 kcal (168.2 kJ) per mole.

If nitric acid loses oxygen to form nitrous acid, it costs the difference between the two atomization energies, 376.1 - 303.1 = 73.0 kcal (305.4 kJ) per mole. Since the N to terminal O energy in HNO_3 is about 104 kcal (435 kJ), the reorganization to HNO_2 releases about 31 kcal (130 kJ)

Table 13:12

Bond Dissociation and Reorganizational Energies of
T4 Halides (IV–I)

cmpd	CBE exp	cmpd	CBE calc	CBE exp	E_R
TiF_4	139.7	TiF	139.5	128–144	0.2
$TiCl_4$	102.7	TiCl	107.6	104.4, 118	−4.9
$TiBr_4$	87.7	TiBr	94.5	88.2, 105	−6.8
TiI_4	70.2	TiI	79.4	64–84	−9.2
ZrF_4	155.3	ZrF	153.0	144.6	2.3
$ZrCl_4$	117.4	ZrCl	120.2	125.4	−2.8
$ZrBr_4$	101.6	ZrBr	106.6	100.2	−5.0
ZrI_4	83.5	ZrI	90.9		−7.4
HfF_4	155.7	HfF	153.3		2.4
$HfCl_4$	118.9	HfCl	121.5		−2.6

per mole. This particular example is hypothetical since the actual decomposition of nitric acid is undoubtedly much more complex. However, solid alkali metal nitrates do decompose to form nitrites.

Release of an oxygen atom from hydrogen peroxide, HOOH, cannot occur simply but overall it forms H_2O. The difference in atomization energy is 255.9 − 221.6 = 34.3 kcal (143.5 kJ), so this is what release of one oxygen atom requires.

Liberation of oxygen from CO_2 has already been shown to cost 127 kcal (531.4 kJ) per mole. CO_2 is capable of acting as an oxidizing agent toward strong reducing agents. Indeed, the most reactive organometallic compounds are spontaneously inflammable in it.

Sulfur trioxide is also oxidizing, and in becoming SO_2 the SO energy increases by 14.9 kcal (62.3 kJ) per mole of bonds. The release of oxygen atom costs the difference in the atomization energies, 339.5 −256.2 = 83.3 kcal (348.5 kJ) per mole, compared to a CBE of 113.2 kcal (473.6 kJ) in SO_3.

Table 13:13

**Bond Dissociation and Reorganizational Energies of
T4 Halides (III–II–I)**

cmpd	BE calc	cmpd	BE calc	E_R	cmpd	BE calc	E_R
TiF_3	151.2	TiF_2	150.6	0.6	TiF	139.5	11.1
$TiCl_3$	110.8	$TiCl_2$	117.0	−6.2	TiCl	107.6	9.4
$TiBr_3$	94.5	$TiBr_2$	103.3	−8.8	TiBr	94.5	8.8
TiI_3	75.4	TiI_2	87.4	−12.0	TiI	79.4	8.0
ZrF_3	168.2	ZrF_2	164.8	3.4	ZrF	153.0	11.8
$ZrCl_3$	126.9	$ZrCl_2$	130.3	−3.4	ZrCl	120.2	10.1
$ZrBr_3$	109.6	$ZrBr_2$	116.0	−6.4	ZrBr	106.6	9.4
ZrI_3	89.9	ZrI_2	99.5	−9.6	ZrI	90.9	8.6
HfF_3	168.6	HfF_2	165.2	3.4	HfF	153.3	11.9
$HfCl_3$	128.5	$HfCl_2$	131.7	−3.2	HfCl	121.5	10.2

In H_2SO_4, the release of one oxygen atom would presumably form SO_2 and H_2O. Therefore the energy required can be determined from the atomization energies: 585.6 − (256.2 + 221.6) = 107.8 kcal (451.0 kJ) per mole.

Release of oxygen atom from HOCl would allow formation of HCl, so the energy required is the difference between the atomization energies: 162.7 − 103.2 = 59.5 kcal (248.9 kJ) per mole.

HALOGENATION BY HALIDES

It is of interest to examine similarly some properties of oxyhalides. For example, phosgene, $COCl_2$, is known as a good high temperature chlorinating agent. Its atomization energy is 341.2 kcal (1427.6 kJ), which includes 2 C–Cl bonds totalling 174.6 kcal (730.5 kJ), leaving 166.6 kcal (697.1 kJ) for the CO bond. Release of two Cl atoms, however, allows the reorganization of the CO portion to carbon monoxide molecules with bond energy 257 kcal per mole. The E_R value is then 167 − 257 = −90 kcal (−376.6 kJ). This means that

Table 13:14

Bond Dissociation Energies of T4 Halides

TiF_4	128	TiF_3	152	TiF_2	162	TiF
$TiCl_4$	95	$TiCl_3$	105	$TiCl_2$	126	TiCl
$TiBr_4$	81	$TiBr_3$	86	$TiBr_2$	112	TiBr
TiI_4	65	TiI_3	63	TiI_2	95	TiI
ZrF_4	142	ZrF_3	172	ZrF_2	177	ZrF
$ZrCl_4$	108	$ZrCl_3$	124	$ZrCl_2$	140	ZrCl
$ZrBr_4$	94	$ZrBr_3$	103	$ZrBr_2$	125	ZrBr
ZrI_4	77	ZrI_3	80	ZrI_2	108	ZrI
HfF_4	143	HfF_3	172	HfF_2	177	HfF
$HfCl_4$	109	$HfCl_3$	125	$HfCl_2$	142	HfCl

release of the chlorine atoms will require not the 174.6 kcal (730.5 kJ) of the two CBE's, but only 84.6 kcal, an average of 42.3 kcal (177.0 kJ) per mole apiece.

A more reactive chlorinating agent is sulfuryl chloride, SO_2Cl_2. Its atomization energy is 330.3 kcal (1382 kJ) per mole, which includes two S-Cl bonds at 69.9 kcal (292.5 kJ) each. This leaves 330.3 - 2 x 69.9 = 190.5 kcal (797.1 kJ) for the two SO bonds. Once the chlorine has gone, however, the residue can revert to SO_2, with atomization energy 256.2 kcal (1071.9 kJ). The total E_R for removal of two chlorine atoms is therefore 190.5 - 256.2 = -65.7 kcal (-274.9 kJ). Removal of both Cl atoms requires only 74.1 or 37 kcal (154.8 kJ) per mole average for each. Sulfuryl chloride is nearly completely decomposed at 100°C.

What about thionyl chloride, $SOCl_2$? Its atomization energy is 234.6 kcal (981.6 kJ) per mole, which includes two S-Cl bonds at 65.6 kcal (274.5 kJ) each. This leaves 103.4 kcal (432.6 kJ) for the SO bond. The energy of this bond in SO molecule, formed if two Cl atoms are released, is 124.7 kcal (521.7 kJ). The total E_R for release of two Cl atoms is therefore 103.4 - 124.7 = -21.3 or an average of -10.7 kcal (-44.8 kJ) per mole of Cl atoms. The average BDE is then

Table 13:15

Relative Ease of Release of Atomic Oxygen

O_3	25.5	HOCl	59.5	SO_3	83.3
H_2O_2	34.3	HNO_3	73.0	H_2SO_4	107.8
N_2O	40.2	NO_2	73.5	CO_2	127.0

65.6 - 10.7 = 54.9 kcal (229.7 kJ) per mole, much higher than in SO_2Cl_2.

Similarly, the possible fluorinating ability of SO_2F_2 and SOF_2 may be studied. The atomization energy of SO_2F_2 is 428.1 kcal (1791.2 kJ). This includes two S-F bonds at 99.8 kcal each, or 199.6 kcal (835.1 kJ). Subtracting these from the atomization energy leaves 228.4 kcal (955.6 kJ) for the SO_2 fragment. This can reorganize forming SO_2 molecule with atomization energy of 256.2 kcal (1071.9 kJ), so E_R is 228.4 - 256.2 = -27.8 kcal (-116.3 kJ) per mole. This would reduce the BDE's for the S-F bonds to 199.6 - 27.8 = 171.8 kcal (718.8 kJ) or an average of 85.9 kcal (359.4 kJ) per mole of S-F bonds. SO_2F_2 would not seem very promising as a fluorinating agent at ordinary temperatures, being too stable. In fact it is quite inert and unreactive.

The atomization energy of SOF_2 is 276.6 kcal (1157.3 kJ), to which the S-F bonds contribute 83.4 kcal apiece or 166.8 kcal (697.9 kJ). This leaves 109.8 kcal (459.4 kJ) for the SO bond. In SO molecule, the bond energy is 124.7 kcal (521.7 kJ). E_R for breaking off two fluorine atoms is therefore -14.9 kcal (-62.3 kJ). Subtracting this from 166.8 leaves 151.9 or 78 kcal (326.4 kJ) for the average S-F bond, still too high for easy loss at ordinary temperatures. Thionyl fluoride is a stable, unreactive gas.

Sulfuryl bromide is apparently too unstable to exist, but thionyl bromide, $SOBr_2$, is known. Its atomization energy is 201.1 kcal (841.4 kJ), to which the two S-Br bonds contribute a total of 111.2 kcal (465.3 kJ). This leaves only 89.9 kcal (376.1 kJ) for the SO bond, which in the SO molecule is 124.7. E_R is -34.8, which subtracted from 111.2 = 76.4 kcal (319.7 kJ). An average 38.2 kcal suffices to break loose the Br atoms. Thionyl bromide decomposes below its boiling point.

Chapter 14

BOND ENERGY IN NONMOLECULAR SOLIDS

The application of the theory of polar covalence to determining and understanding the bond energy in both molecular and nonmolecules **solids** encounters problems not present in the treatment of gaseous molecules. These arise from the proximity of atoms not part of the same molecule, in molecular solids, and from the influence of closer packing of atoms, in nonmolecular solids. The emphasis in this chapter will be on nonmolecular solids, and especially on those conventionally termed "ionic."

THE IONIC MODEL

The familiar model of such a solid is the "ionic" model, developed by Born, Mayer, and others. Therein the individual atoms are imagined to be ions resembling hard spheres, packed together so that each cation is surrounded by anions and each anion surrounded by cations. The charges are imagined as centered at the nuclei of the ions, so that the model represents also a collection of point charges. The attractions between opposite charges are considered to provide the binding force.

These attractions are not, however, confined to adjacent neighbors, but extend throughout the crystal. Each ion is surrounded by attractive oppositely charged ions, but just beyond this immediate sphere lie repulsive ions of like charge, and beyond them, attractive ions again of unlike charge, and so throughout the crystal. The point charge picture allows the calculation of the net attractive forces, when all the attractions and repulsions are summed, provided the crystal structure is known. It is found that the net attraction is equal to the attraction between one pair of oppositely charged ions, multiplied by a structural constant known as the Madelung constant. Madelung constants for various types of crystal structure are presented in Table 14:1.

A collection of point charges, however, would obviously be unstable, as there would be nothing to prevent collapse toward the center of the crystal. In fact, such collapse is prevented by the existence of an electronic sphere around each nucleus, whether it be cation or anion. Repulsions

Table 14:1

Madelung Constants

structure type	coordination no.	Madelung constant
NaCl (rock salt)	6:6	1.748
CsCl (cesium chloride)	8:8	1.762
ZnS (zinc blende)	4:4	1.631
ZnS (wurtzite)	4:4	1.641
CaF_2 (fluorite	8:4	5.039
Cu_2O (cuprite)	2:4	4.442
TiO_2 (rutile)	6:3	4.770
SiO_2 (beta-quartz)	4:2	4.402
Al_2O_3 (corundum)	6:4	24.242

between adjacent spheres allow packing of the atoms only to the point where repulsions equal attractions. The Madelung energy is therefore reduced by a factor which may be called the repulsion coefficient. This may be determined from measurements of the compressibility of the crystal but is usually obtained from the Born values of n, where $k = 1 - 1/n$. Values for M8 (inert shell) elements are given in Table 14:2. The "lattice energy," U, which is the energy released when gaseous ions are brought together to form the crystal, is a function of the product of the ionic charges times the Madelung constant times the repulsion coefficient divided by the observed internuclear distance between adjacent oppositely charged ions.

$$U = ze^+ \times ze^- \times M \times k/R_o$$

When R_o is expressed in picometers (pm) and U in kcal per mole, the factor 33200 is employed.

The lattice energy for the solid alkali halides obtained by this calculation agrees quite well with the value obtained by means of a Born-Haber thermochemical cycle. Lattice energy is sometimes considered a measure of the stability of the crystal. Not so, except with respect to ionization in

Table 14:2

Repulsion Coefficients

$(k = 1 - 1/n)$

no. electrons in ion	n	k
2	5	0.80
10	7	0.86
18	9	0.89
36	10	0.90
54	12	0.92

a solvent. The **atomization energy** suggests the stability of the crystal, for it is the total bond energy in the solid. Heating the crystal, unless there is decomposition, releases first gaseous molecules and then free atoms, **not** ions. However, some insist that the gaseous MX molecules are "ion-pairs".

A melted salt is commonly thought to consist of relatively free ions, since it conducts electricity with liberation of the expected elements at the electrodes. However, the electrolysis does not necessarily prove the nature of the melt, just as production of hydrogen by electrolysis of a solution of NaOH does not necessarily prove the presence of significant concentration of H^+ ion. Crystal lattice energy is of course of interest in studying the solubility of the solid where it dissolves in the form of solvated ions, but the decomposition of solid to gaseous ions is only an imaginary process. The reason is that electron affinities are never as large as ionization energies. In other words, the transfer of electrons from metal to nonmetal is always endothermic. This has often led to the peculiar reasoning that although a single pair of ions cannot form, for example, from an atom of sodium and an atom of chlorine, the increase in energy resulting from condensation of these molecules to a nonmolecular solid is adequate to bring about the transfer. This reasoning leads to very strained arguments, to the effect that, for example, it is all right to have discrete ions in such a solid as MgO, in which each oxygen, which holds two electrons, one highly endothermically, is surrounded by six magnesium ions which

exert a powerful attraction for electrons as measured by
the sum of the first two ionization energies (526 kcal, 2201
kJ), and each electron-hungry magnesium ion is similarly
surrounded by electron-releasing oxide ions, and nothing happens. The ions are "stabilized by the crystal energy." It
seems little more logical to expect ions to refrain from interacting under such conditions than it would to expect sodium
chloride to consist of sodium atoms surrounded by electron-
hungry chlorine atoms and chlorine atoms to be surrounded
by sodium atoms, without any interaction.

Consider the nature of any simple cation. It is a nucleus
imbedded in a cloud of electrons, insufficient to balance
the nuclear charge. It is an atom having outer vacancies
within which the effective nuclear charge is substantial because
the residual electrons are insufficient to cover it up. It is
a potential electron pair acceptor.

Any simple anion is also a nucleus imbedded within
a cloud of electrons, but here the electrons are more than
sufficient to balance and block off the nuclear charge. The
exterior of the anion consists of pairs of electrons. Their
potential availability for donation is enhanced by the negative
charge on the ion.

When a solid is described as consisting of discrete
ions, as in the ionic model, this suggests that a close packed
arrangement is stable in which potential electron pair donors
are completely surrounded by potential electron pair acceptors
and acceptors by donors, and no interaction occurs. It seems
much more probable that there is some degree of coordination
between anions and cations such that the individual ions lose
their identity as discrete ions. An appreciable covalent contribution becomes part of the total atomization energy.
Consistent with this is the observation from careful X-ray
analyses that a plot of electron density along the bond axis
always shows a minimum, but this is never at the expected
"ionic" junction. It always suggests a shorter radius for the
"anion" and a larger radius for the "cation", consistent with
only partial charges on these atoms.

THE COORDINATED POLYMERIC MODEL

From the viewpoint of electronegativity equalization,

there is no combination of metal with nonmetal in which equalization does not occur before, and often long before, complete transfer of valence electrons occurs. Let us then consider a "coordinated polymeric" model as an alternative to the "ionic" model. The ionic model will not be abandoned completely; in fact, it is indispensable for the evaluation of the ionic contribution to the total bond energy in a non-molecular solid. However, the theory of polar covalence can operate very well for such solids, provided appropriate modifications are made.   It is assumed that the same state of a nonmolecular solid can be reached in either of two directions. (1) It may be formed by interaction of the free elements through the equalization of electronegativity. (2) It may be imagined as formed by condensation of gaseous cations and anions, forming a solid structure in which an appreciable degree of coordination introduces a significant covalent character to the bonding. This results in the equalization of electronegativity and a loss of identity of the atoms as true ions.

To determine the covalent contribution to the total bond energy in a solid, a major change in the bonding electrons must be recognized. Instead of just the normal bonding electrons that would be involved in the gaseous molecule, now all the vacant orbitals of the metal atom are potentially available and all the electrons of the outer principal quantum level of the nonmetal likewise. Therefore the usual expression for covalent energy, the geometric mean of the two homonuclear single covalent bond energies corrected for any difference between covalent radius sum and actual bond length, must be multiplied by a factor n, usually 4 but for reasons not yet understood, apparently only 3 in halides of Li, Na, Be, and Mg. For the ionic energy contribution, the usual expression for lattice energy is used. The total atomization is then:

$$E = t_c E_c + t_i E_i = t_c n R_c (E_{MM} E_{XX})^{1/2}/R_o \\ + 33200 t_i z e^+ z e^- Mk/R_o$$

One more difference is in the evaluation of t_i. Whereas in molecules it is half the difference between the charges, in nonmolecular solids it is numerically equal to the partial charge on any univalent atom if present. For example, the partial charges in $MgCl_2$ are Mg 0.658 and Cl -0.329. The value of t_i is 0.493 for the gas but 0.329 for the solid.

In Chapter 3, the bond energy in a gaseous molecule of KBr was calculated to illustrate the application of the theory of polar covalence to molecules. It was found that this bond energy is the sum of a covalent contribution of 7.6 kcal (31.8 kJ) and an ionic contribution of 84.5 kcal (353.5 kJ) per mole, for a total of 92.1 kcal (385.3 kJ) compared to the experimental value of about 91.5 kcal (382.8 kJ). In the crystal, the bond length changes from 282.1 to 328.5 pm. The covalent contribution is then

$$t_c E_c = 0.282 \times 4 \times 24.6 \times 310.4/328.5 = 26.2 \text{ kcal } (109.6 \text{ kJ})$$

For calculation of the ionic contribution, the Madelung constant of 1.748 and the repulsion coefficient of 0.89 must be used:

$$t_i E_i = 0.718 \times 33200 \times 1.748 \times 0.89/328.5 = 112.7 \text{ kcal}$$

The total calculated atomization energy is the sum, 138.9 kcal (581.2 kJ), which is practically identical with the experimental value of 138.6 kcal (579.9 kJ).

Consistent with the idea of the coordinated polymeric model, the covalent contribution in the nonmolecular solid is about 19 per cent of the total energy, compared with only 8 per cent in the gaseous molecule. The difference in total energy, 138.6 - 91.5 = 47.1 kcal (197.1 kJ), provides the incentive for the condensation of KBr gas molecules, amply overcoming the decrease in entropy that results from crystalization.

The results of applying this coordinated polymeric model to the calculation of atomization energies of the alkali metal and alkaline earth metal halides are given in Tables 13:3 and 13:4. It will be seen that the agreement between calculated and exxperimental values is in general excellent, better than could be obtained from the ionic model.

In application to the oxides and sulfides of these elements, it appeared that the ionic contributions were always too high, giving calculated atomization energies much higher than the experimental values. It seemed likely that the compressibility of anions of double negative charge might be much greater than of unit charged ions, and that a correction of the repulsion coefficient was required. Empirically it

Table 14:3

Atomization of Solid Alkali Metal Halides

cmpd	t_i	R_o	k	M	E_c	E_i	Ecalc	Eexp
LiF	0.753	200.9	0.83	1.748	23.5	180.4	204.1	204.2
LiCl	0.666	257.	0.86	1.748	34.7	129.3	163.6	164.8
LiBr	0.621	275.	0.87	1.748	34.5	114.0	148.5	148.7
LiI	0.540	300.2	0.88	1.748	36.2	91.0	128.5	128.2
NaF	0.797	231.	0.86	1.748	15.1	172.2	187.3	181.9
NaCl	0.711	281.4	0.88	1.748	24.1	129.2	153.0	153.0
NaBr	0.666	298.0	0.88	1.748	24.8	114.1	139.0	138.7
NaI	0.585	323.1	0.89	1.748	26.9	93.5	120.4	120.0
KF	0.849	266.4	0.87	1.748	13.8	161.1	174.7	175.8
KCl	0.762	313.8	0.89	1.748	24.7	125.3	150.0	154.6
KBr	0.718	328.5	0.89	1.748	26.2	112.7	138.9	138.6
KI	0.637	352.5	0.90	1.748	29.4	94.3	123.7	125.1
RbF	0.918	282.	0.87	1.748	7.4	164.4	171.8	171.5
RbCl	0.831	326.7	0.89	1.748	17.5	131.4	148.9	152.4
RbBr	0.787	341.8	0.90	1.748	19.6	119.8	139.4	140.8
RbI	0.706	365.5	0.91	1.748	23.8	101.9	125.7	124.6
CsF	0.975	300.4	0.89	1.748	2.1	167.9	170.0	171.5
CsCl	0.889	356.6	0.90	1.763	10.4	131.5	141.9	153.1
CsBr	0.844	371.3	0.91	1.763	13.1	120.8	133.9	141.9
CsI	0.763	394.7	0.92	1.763	18.2	104.1	122.3	126.5

was found that correction factors of 0.644 for oxides, 0.732 for M1 sulfides, and 0.615 for M2 sulfides were appropriate. Results of applying these factors in the calculation of atomization energies of M1 and M2 oxides and sulfides are given in Table 14:5. It will be observed that the results are not as good as for the halides, but of the correct order of magnitude. Obviously there is much room for improvement, both in accuracy and in understanding.

The coordinated polymeric model blends well with the concept of dissolution through solvation of ions. In the solid, the atoms may be viewed as ions each within a coordination sphere of oppositely charged ions. In dissolving, these ions merely change the nature of their close coordination

Table 14:4

Atomization of Solid Alkaline Earth Metal Halides

cmpd	t_i	R_O	k	M	E_c	E_i	Ecalc	Eexp
MgF_2	0.394	202.	0.84	4.76	102.0	258.9	360.9	356.7
$MgCl_2$	0.329	254.	0.87	4.49	93.0	167.5	260.5	261.7
$MgBr_2$	0.295	270.	0.88	4.38	87.1	139.8	226.5	229.1
MgI_2	0.234	294.	0.89	4.38	82.7	103.0	185.7	188.4
CaF_2	0.486	236.	0.87	5.04	74.5	299.8	374.3	372.6
$CaCl_2$	0.418	273.	0.89	4.27?	98.3	192.5	290.8	291.5
$CaBr_2$	0.383	294.	0.89	4.38?	91.2	168.6	259.8	259.9
CaI_2	0.320	304.	0.90	4.38	91.6	137.8	229.4	224.7
SrF_2	0.554	251.	0.87	4.82	58.2	307.3	365.5	368.3
$SrCl_2$	0.484	302.	0.90	4.62	75.0	221.2	296.2	295.9
$SrBr_2$	0.449	321.	0.90	4.62	70.6	193.1	263.7	264.7
SrI_2	0.384	342.	0.91	4.62	69.6	156.7	226.3	224.2
BaF_2	0.578	268.	0.88	5.04	50.3	317.6	367.9	367.4
$BaCl_2$	0.508	318.	0.90	5.04	66.1	240.6	306.7	304.6
$BaBr_2$	0.472	338.	0.91	5.04	62.4	212.6	275.0	275.8
BaI_2	0.407	359.	0.91	5.04	61.9	172.6	234.6	236.3

spheres from opposite ions to polar molecules. In this sense, the process of dissolution in a solvent does not differ greatly from changes in coordination within a solvent when a ligand more effective than the solvent is added.

However, the nature of liquefaction through the agency of a solvent can be much more complex than that. Unfortunately, standard tabulations of solubilities usually express solubilities in grams of solute per 100 grams of solvent. If these values are converted to the far more meaningful moles of solute per mole of solvent, it becomes obvious that there are many very soluble salts which in their saturated solutions have far too few solvent molecules to surround either cation or anion. Clearly the usual picture of ions imbedded in spheres of solvent is wholly inadequate for such concentrated solutions. Very interesting problems for future research are presented.

Table 14:5

Atomization of M1 and M2 Oxides and Sulfides

cmpd	t_i	R_o	k	M	E_c	E_i	Ecalc	Eexp
Li_2O	0.396	200.	0.53	5.04	106.0	177.1	283.1	279.7
Li_2S	0.334	248.	0.65	5.04	94.0	141.9	235.9	247.9
Na_2O	0.414	241.	0.54	5.04	51.2	157.3	208.5	210.3
Na_2S	0.353	282.	0.65	5.04	71.3	131.8	203.1	204.7
K_2O	0.432	279.	0.56	5.04	45.6	146.8	192.4	188.8
K_2S	0.374	320.	0.67	5.04	43.2	147.1	190.3	199.9
Rb_2O	0.452	292.	0.56	5.04	43.8	146.8	190.6	179.8
Rb_2S	0.397	332.	0.68	5.04	41.7	150.1	191.8	191.1
Cs_2O	0.464	286.	0.57	4.38	40.8	126.5	167.3	179.6
Cs_2S	0.412	348.	0.68	5.04	38.5	147.0	185.5	188.6
MgO	0.243	210.	0.52	1.748	113.5	138.9	252.4	253.8
MgS	0.182	260.	0.65	1.748	116.2	85.9	202.1	199.5
CaO	0.299	240.	0.54	1.748	91.9	158.3	250.2	254.7
CaS	0.238	284.	0.66	1.748	123.0	104.1	227.1	224.1
SrO	0.339	257.	0.55	1.748	77.2	169.1	246.3	240.9
SrS	0.277	294.	0.66	1.748	106.8	118.4	225.2	218.4
BaO	0.352	276.	0.55	1.748	68.8	164.0	232.8	233.3
BaS	0.291	318.	0.66	1.748	94.3	114.9	209.2	217.6

NONMOLECULAR SOLIDS WITH LESS POLAR BONDS

As the polarity of the bonding decreases, apparently the effects of charged atoms more distant than the nearest neighbors become less important, and the simple theory of polar covalence together with the coordinated polymeric model become less appropriate. The atomization energies are calculated to be appreciably larger for such compounds than the experimental values. In a sense, the atoms in the nonmolecular solid tend to resemble more and more the atoms in the molecules, being less and less influenced by nonbonded neighbors and near neighbors.

Examples of the problems faced in dealing with less polar nonmolecular solids are provided by a group of 1:1 compounds which exhibit 4:4 coordination. These include one tetravalent, eleven trivalent, thirteen divalent, and 5 monovalent compounds. The essential data and calculated bond energies are summarized in Table 14:6. One might expect the four bonds formed by each atom to be an average of regular polar covalent and coordination bonds according to the valence. If these bonds were all equivalent to normal polar covalent bonds, the atomization energy would be four times the calculated bond energy. It never is that large, as shown by the tabulated ratios of experimental atomization energy to calculated bond energy. The ratios are quite variable, and the only observable generalization that can be made is that the ratios average 3.16 for the trivalent atoms, 2.38 for the divalent, and 2.14 for the monovalent. In other words, the tendency is toward closer approach to 4, the more regular polar covalent bonds can be formed.

First an explanation must be sought for the SiC ratio of 3.55 instead of 4. Assuming the experimental value to be correct, it seems likely that steric repulsions among the four Si atoms crowded around one C atom weaken the bonding. It may be recalled that the atomization energy of CCl_4 is substantially less than calculated, apparently for this reason. The negative charge on carbon and the positive charge on silicon would reduce the radius ratio somewhat, but it is still comparable to that in CCl_4. Whether the opposite polarity would make a difference is not known.

Next, the theoretical number of normal bonds may be multiplied by the calculated bond energy and subtracted from the experimental atomization energy to see how much is left for coordination energy. Here is encountered the problem that in AlP, AlAs, InN, and InAs, the atomization energy is too low to account even for three normal covalent bonds. Steric weakening might account for the first three but hardly for AlP. In all the other compounds the residue of energy for coordination per bond is much smaller than the calculated bond energy. It can only be concluded that coordination energy contributes relatively little per bond compared to the polar covalence. The roles of steric repulsions and atomic charges are probably very significant but difficult to evaluate.

Table 14:6

Bond Energies in 1:1 Solids Showing 4:4 Coordination

cmpd	R_O	t_i	E_c	E_i	Ecalc	at. E	At E/E
SiC	189.	0.124	61.2	21.8	83.0	294.9	3.55
AlN	186.	0.305	32.2	54.4	86.6	267.8	3.09
AlP	235.	0.176	42.2	24.9	67.1	198.0	2.95
AlAs	243.	0.235	34.3	32.1	66.4	178.9	2.69
GaN	194.	0.148	34.5	25.3	59.8	204.8	3.42
GaP	236.	0.019	45.7	2.7	48.4	165.8	3.43
GaAs	243.	0.078	37.7	10.7	48.4	154.7	3.20
GaSb	265.	0.008	36.1	1.0	37.1	138.6	3.74
InN	213.	0.207	26.2	32.3	61.5	175.2	2.85
InP	255.	0.079	39.2	10.3	49.5	158.6	3.20
InAs	261.	0.137	32.1	17.4	49.5	144.3	2.92
InSb	282.	0.067	31.3	7.9	39.2	128.5	3.28
BeO	164.	0.361	28.4	73.1	101.5	282.8	2.79
BeS	210.	0.239	42.7	37.8	80.5	199.7	2.48
MgTe	275.	0.299	27.7	36.1	63.8	132.	2.07
ZnO	195.	0.268	24.9	45.6	70.5	174.0	2.47
ZnS	233.	0.146	38.0	20.8	58.8	143.4	2.44
ZnSe	245.	0.156	31.3	21.1	52.4	124.5	2.38
ZnTe	263.	0.081	33.8	10.2	44.0	106.3	2.42
CdO	235.	0.322	19.5	45.5	65.0	148.0	2.28
CdS	252.	0.199	33.0	26.2	59.2	131.6	2.22
CdTe	278.	0.135	29.8	16.1	45.9	95.8	2.09
HgS	253.	0.151	18.6	19.8	38.4	94.8	2.47
HgSe	263.	0.162	15.4	20.5	35.9	80.0	2.23
HgTe	279.	0.087	16.8	10.4	27.2	71.7	2.64
CuF	185.	0.366	24.5	65.7	90.2	159.	1.76
CuCl	235.	0.279	30.4	39.4	69.8	142.7	2.04
CuBr	246.	0.235	29.2	31.7	60.9	132.8	2.18
AgI	280.	0.201	25.8	23.8	49.6	108.3	2.18

MOLECULAR ADDITION COMPOUNDS

It is often stated that a coordinate covalent bond differs from a normal covalent bond in that one atom provides both electrons and the other, both vacancies, but once formed, is indistinguishable from ordinary covalence. The latter may be true in such complexes as H_3O^+ and NH_4^+ and substituted ammonium ions, but is probably untrue in most complexes, including molecular addition compounds.

Let us consider, for example, the gaseous dimers of aluminum halides. Here two AlX_3 molecules combine so that two AlX_2 groups are held together in one plane by two X bridges above and below at right angles to the plane. Calculations show that the bonds in the monomers are accurately represented as Al-X'''. If it is assumed that these change to Al-X' with dimerization, and the bridge bonds are normal polar covalent bonds, the calculated atomization energy of the dimer is far too high. The average bridge bond energy obtained by difference is as follows: Al_2F_6 80.1 (calc 115.1) kcal (335.1 kJ); Al_2Cl_6 67.1 (calc 87.7 kcal (280.7 kJ); Al_2Br_6 53.4 (calc 79.8) kcal (223.4 kJ); and Al_2I_6 44.4 (calc 63.0) kcal (185.8 kJ) per mole of bonds. The experimental energies of dimerization are, Al_2F_6 -52.2 (-218.4 kJ), Al_2Cl_6 -29.7 (124.3 kJ), Al_2Br_6 -28.8 (-120.5 kJ), and Al_2I_6 -24.3 (101.7 kJ) kcal per mole.

There must be, of course, some energy loss in the change from planar to tetrahedral configuration, but this cannot easily be calculated. It seems quite unlikely that this could account for seemingly weaker bridging, because bond energy calculations appear to be accurate regardless of structure, assuming no steric interference.

Data are available for a number of other molecular addition compounds but they have not yet revealed their secret. A detailed discussion would not be very useful at this time. Perhaps here the complexities require more than a simple approach. As observed years ago, electronegativity equalization does not necessarily occur in all coordination, and perhaps especially where the donor is more electronegative than the acceptor.

In summary, the modified theory of polar covalence is quantitatively accurate in application to highly polar nonmolecular solids, but to date meets obstacles where the polarity is not very high. No matter how logical the approach to research may seem in advance, the unpredictability of its results makes practically certain a sometimes embarrassingly illogical sequence of discovery, which can only be rearranged and made orderly in retrospect. Ideally, the next major step in this bond energy work would lead to enlightenment about the quantitative nature of coordination. Perhaps, as suggested some years ago, the capacity of a ligand to serve as an electron reservoir may have as much to do with its electron pair donating ability as the availability of lone pair electrons and the partial charge on the atom providing them.

If this is so, the problem may prove far more complex than that of describing simple polar covalence.

Chapter 15

THE SIGNIFICANCE OF PARTIAL CHARGE IN UNDERSTANDING PHYSICAL AND CHEMICAL PROPERTIES

Healthy controversy, in my opinion, is essential to progress in science or any other area of human knowledge. Yet, unfortunately, there seems to be a prevailing attitude that expression of dissension is somehow unseemly, that a scientist has no business being argumentative. He should simply state the facts and let others decide for themselves. But what if these facts, although proven uniquely useful, are scorned or ignored? Should their discoverer remain cool and silent, meekly accepting the unwarranted rejection of his contribution, or should he keep trying? The ever accelerating flood of new scientific literature, completely impossible for any individual to assimilate, creates a real danger, it seems to me, that valuable ideas may be swept unseen or unappreciated downstream too rapidly for adequate evaluation, and lost, if not forever, for too long.

The determination of partial charge based on electronegativity equalization was first described more than a quarter of a century ago (B13–B16). The use of this charge as an index of the condition of combined atoms, which correlates closely with physical and chemical properties, has been documented in great detail (B13–B18, B1, B2). This was long before the application of partial charge to determining the energy of polar covalent bonds was even anticipated. Although this work attracted considerable interest initially, the universal preoccupation with quantum mechanics appears to have discouraged acceptance of such simplified alternatives. To this day, despite irrefutable evidence of the successful and indispensable application of partial charge to the calculation of the energy of thousands of polar covalent bonds, as provided herein as well as since 1966 (B27–B34), and despite the relatively recent quantum mechanical verification (A8, A9) of the principle of electronegativity equalization, other authors of textbooks of general and inorganic chemistry and many others as well refuse to recognize the validity and usefulness of these concepts.

This refusal would be understandable only if some scientific explanation were provided. No such explanation

179

has ever been given. The excuse, if any, seems to be, "some years have passed since these ideas were proposed and they have not found widespread acceptance," as though nonacceptance were a sound scientific basis for nonacceptance. Mostly, however, this work has been ignored as though it never existed. Such attitudes have resulted in the consistent withholding of financial support of this work, and repeated opinionated censorship with no editorial insistence that reviewers provide scientific justification of their recommendations. Far worse, for many years it has deprived students (except my own) of an ease and depth of understanding of chemistry that they could easily have acquired.

I am so thoroughly convinced that these concepts and applications contribute importantly to an understanding of chemistry that I am impelled to reemphasize the facts that support the validity of the concept of partial charge. In addition to the thousands of bond energies themselves, presented throughout this book, I shall provide more details of the **independent evaluation of partial charge,** the use of such charges **to estimate radii of combined atoms,** and finally the general applicability to **understanding chemical properties.**

PARTIAL CHARGES FROM EXPERIMENTAL DATA

Earlier in this book it was pointed out that partial charge can be evaluated by a method entirely independent of any but experimentally verifiable quantities, with no reliance whatever on electronegativity, equalization, or any related assumptions. In Table 15:1 are presented 191 examples, for which the average difference between partial charge as determined from electronegativities and as calculated independently from experimental data is only 0.007 charge unit. Notice that a complete range of charge values from +1 to -1 is covered. Can this agreement be dismissed as contrived or coincidental? How much evidence must there be?

There have been many other sets of electronegativity values proposed, derived in a wide variety of ways, but the **only real test** of their validity has been the extent of their agreement with earlier accepted values, **which themselves have never been quantitatively proven.** Also there have been many quantum mechanical evaluations of atomic charges, many of which are so logically and chemically unreasonable

Table 15:1

Comparison of Partial Charges from (A) Electronegativity and (B(Independent Experimental Data

	A	B		A	B		A	B
BaF_2c	1.16	1.16	$AlBr_3g$	0.50	0.49	CS_2g	0.05	0.05
SrF_2c	1.10	1.12	MgI_2g	0.47	0.46	H_2Sg	0.05	0.05
$BaCl_2c$	1.02	1.00	ZnF_2g	0.46	0.47	H_2Seg	0.05	0.05
$CsFc$	0.98	0.97	SbF_3g	0.44	0.45	SbH_3g	0.04	0.04
CaF_2c	0.97	0.96	$SiCl_4g$	0.44	0.47	CSg	0.04	0.03
CaF_2g	0.97	0.97	$InCl_3g$	0.41	0.44	Cl_2Og	0.02	0.02
$SrCl_2c$	0.96	0.96	GeF_4g	0.40	0.44	PH_3g	0.02	0.02
$BaBr_2c$	0.94	0.94	SiO_2c	0.40	0.40	AsH_3g	0.02	0.02
$RbFc$	0.92	0.92	SiO_2g	0.40	0.41	CH_4g	0.01	0.01
$SrBr_2c$	0.90	0.90	$BeBr_2g$	0.40	0.40	H_2Teg	0.00	0.00
$CsClc$	0.89	0.89	$PbCl_4g$	0.38	0.35	GeH_4g	0.00	0.00
KFc	0.85	0.86	CF_4g	0.37	0.37	PH_3g	-0.01	-0.01
$CsBrc$	0.84	0.84	AlI_3g	0.36	0.33	GeH_4g	-0.01	-0.01
$NaFc$	0.80	0.76	$HgCl_2g$	0.34	0.34	SbH_3g	-0.01	-0.01
$RbBrc$	0.79	0.80	$BiCl_3g$	0.34	0.33	H_2Teg	-0.01	-0.01
$RbBrg$	0.79	0.77	$InBr_3g$	0.33	0.35	CS_2g	-0.03	-0.02
MgF_2g	0.79	0.80	$ZnCl_2g$	0.33	0.33	CSg	-0.04	-0.03
MgF_2c	0.79	0.78	$SiOg$	0.29	0.28	BiI_3g	-0.04	-0.04
$CaBr_2c$	0.76	0.76	BBr_3g	0.29	0.27	Cl_2Og	-0.04	-0.05
$KClg$	0.76	0.76	PCl_3g	0.28	0.28	CH_4g	-0.05	-0.05
SrI_2c	0.76	0.76	$GeCl_4g$	0.27	0.29	NO_2g	-0.05	-0.06
$CsIg$	0.76	0.76	$ZnBr_2g$	0.27	0.27	BI_3g	-0.05	-0.05
$LiFc$	0.75	0.75	$HgBr_2g$	0.27	0.27	SO_3g	-0.06	-0.06
$KBrc$	0.72	0.71	P_4O_6g	0.25	0.25	AsH_3g	-0.07	-0.07
$KBrg$	0.72	0.71	HFg	0.25	0.25	NF_3g	-0.07	-0.06
$NaClc$	0.71	0.71	CO_2g	0.22	0.22	InI_3g	-0.07	-0.08
$RbIc$	0.71	0.69	NF_3g	0.21	0.19	$GeCl_4g$	-0.07	-0.07
$RbIg$	0.71	0.70	PBr_3g	0.20	0.19	PBr_3g	-0.07	-0.06
$LiClc$	0.67	0.68	InI_3g	0.20	0.23	SO_2g	-0.08	-0.08
$NaBrc$	0.67	0.66	SO_3g	0.19	0.19	NOg	-0.08	-0.08
KIc	0.64	0.66	SO_2g	0.17	0.17	ZnI_2g	-0.08	-0.07
KIg	0.64	0.65	ZnI_2g	0.16	0.14	PCl_3g	-0.09	-0.09
$LiBrc$	0.62	0.62	BI_3g	0.16	0.15	H_2Sg	-0.09	-0.09
BeF_2g	0.60	0.61	$SiSg$	0.16	0.17	CF_4g	-0.09	-0.09
$MgBr_2g$	0.60	0.61	$HClg$	0.16	0.16	GeF_4g	-0.10	-0.11
SiF_4g	0.60	0.64	BiI_3g	0.16	0.15	BBr_3g	-0.10	-0.09
$NaIc$	0.59	0.58	SOg	0.12	0.14	$PbCl_4g$	-0.10	-0.09
$AlCl_3g$	0.58	0.58	$HBrg$	0.12	0.12	CO_2g	-0.11	-0.11
PbF_4g	0.54	0.56	H_2Og	0.12	0.12	H_2Seg	-0.11	-0.11
$LiIc$	0.54	0.55	NO_2g	0.11	0.12	$SiCl_4g$	-0.11	-0.12
BF_3g	0.51	0.48	NOg	0.08	0.08	$BiCl_3g$	-0.11	-0.11
			NH_3g	0.06	0.06	$InBr_3g$	-0.11	-0.12

Table 15:1 cont.

	A	B		A	B		A	B
AlI$_3$g	-0.12	-0.11	ZnF$_2$g	-0.23	-0.24	BaF$_2$c	-0.58	-0.58
SOg	-0.12	-0.14	HFg	-0.25	-0.25	NaIc	-0.59	-0.58
HBrg	-0.12	-0.12	H$_2$Og	-0.25	-0.25	LiBrc	-0.62	-0.62
ZnBr$_2$g	-0.13	-0.13	SiOg	-0.29	-0.28	KIc	-0.64	-0.66
PbF$_4$g	-0.13	-0.14	BeF$_2$g	-0.30	-0.31	KIg	-0.64	-0.65
InCl$_3$g	-0.14	-0.15	MgBr$_2$g	-0.30	-0.30	LiClc	-0.67	-0.68
HgBr$_2$g	-0.14	-0.13	CaI$_2$c	-0.32	-0.30	RbIc	-0.71	-0.69
SiF$_4$g	-0.15	-0.16	SrI$_2$c	-0.38	-0.38	RbIg	-0.71	-0.70
SbF$_3$g	-0.15	-0.15	SrI$_2$g	-0.38	-0.39	NaClc	-0.71	-0.71
SiSg	-0.16	-0.17	CaBr$_2$c	-0.38	-0.38	KBrc	-0.72	-0.71
HClg	-0.16	-0.16	MgF$_2$c	-0.39	-0.39	KBrg	-0.72	-0.71
NH$_3$g	-0.17	-0.17	MgF$_2$g	-0.39	-0.40	LiFc	-0.75	-0.75
AlBr$_3$	-0.17	-0.16	BaI$_2$c	-0.41	-0.41	KClg	-0.76	-0.76
P$_4$O$_6$g	-0.17	-0.16	SrBr$_2$c	-0.45	-0.45	CsIg	-0.76	-0.76
ZnCl$_2$g	-0.17	-0.17	BaBr$_2$c	-0.47	-0.47	RbBr c	-0.79	-0.80
HgCl$_2$g	-0.17	-0.17	SrCl$_2$c	-0.48	-0.48	RbBrg	-0.79	-0.77
BF$_3$g	-0.17	-0.16	CaF$_2$c	-0.49	-0.48	CsBrg	-0.84	-0.84
AlCl$_3$g	-0.19	-0.19	CaF$_2$g	-0.49	-0.49	KFc	-0.85	-0.86
SiO$_2$c	-0.20	-0.20	BaCl$_2$c	-0.51	-0.50	CsClg	-0.89	-0.89
SiO$_2$g	-0.20	-0.20	LiIc	-0.54	-0.55	RbFc	-0.92	-0.92
BeBr$_2$g	-0.20	-0.20	SrF$_2$c	-0.55	-0.56	CsFc	-0.98	-0.97
MgI$_2$g	-0.23	-0.23						

that they could never have been accepted for publication if they had not been determined by quantum mechanical approximations. From the literature one might easily gain the impression that the difficulty of obtaining accurate and useful results has led many theoretical chemists to judge their work (and each other's work) solely according to the sophistication and ingenuity of their methods. They seem to have decided, contrary to popular belief, that the proof of the pudding lies not in the eating but in the recipe. It is tempting to discuss some of this work in detail and show how unreal and illogical some of the published charge values are, but I believe the space can be more usefully filled. The matter may be left for the thoughtful consideration of fair-minded readers, with a challenge analogous to that of the automobile salesman: "If you can find a better car, buy it!"

The electronegativity and partial charge values used

**in this work have, to the best of my knowledge, been subjected
to a more thorough testing and evaluation than have any
other values ever proposed,** and they have passed the test.
The calculation of partial charges and their application to
thousands of successful bond energy calculations, bond lengths,
and general chemical interpretations has **not** been achieved
by use of **any other set of electronegativities** or **any other
determinations of atomic charges.**

Nevertheless I would be last to deny the arbitrary
quality of a concept such as that of partial charge. (For
a fuller discussion, see my earlier work (B1, B3)). The
molecular orbital picture of merged atomic orbitals together
with the general wave mechanical picture of electronic clouds
certainly invites a degree of healthy skepticism concerning
the assignment of partial charges to individual atoms in any
combination. There is not, and probably cannot be, any
experimental method of determining or verifying such charges,
since the interpretation of experimental observations is itself
open to variations in reasoning. The evaluation of the concepts
and numbers obtained by their use must therefore be essentially
practical. The system works. I leave it to those dissatisfied
with the simple picture evoked by the theory of polar covalence
to determine more exactly why it works.

THE RADII OF COMBINED ATOMS

The successful application of partial charges to
thousands of accurate bond energy calculations would seem
ample verification of their validity, to which the reverse
calculations just described and summarized in Table 15:1
should provide additional reinforcement. Still another part
of the picture of polar covalence can also be examined, and
this is the radii of atoms in combination. According to the
concept of electronegativity equalization, as an atom acquires
partial negative charge, the electronic cloud expands and
the electronegativity decreases. As an atom acquires partial
positive charge, the electronic cloud contracts and the
electronegativity increases. To what extent can these changes
in atomic radii be determined?

A reasonable assumption worth testing is that the
radius of an atom might change with charge in a roughly

Table 15:2

Values of B for Calculation of Radii of Combined Atoms

$(r = r_0 - B \times \text{charge; gases, normal type, }$ **solids, boldface)**

H	97.2	Cs	117.9	Ba	48.1	F	52.0
		Cs	**94.1**	**Ba**	**35.9**	**F**	**94.1**
Li	115.8	Be	61.2	B	52.6	Cl	68.8
Li	**72.0**					**Cl**	**115.7**
Na	92.5	Mg	62.7	Al	56.5	Br	66.8
Na	**74.4**					**Br**	**116.7**
K	106.0	Ca	63.0	Si	57.3	I	67.0
K	**91.5**					**I**	**128.0**
Rb	112.7	Sr	59.0	P	45.2		
Rb	**100.5**	**Sr**	**44.1**				

linear relation. The radius of a combined atom can then be calculated as equal to the nonpolar covalent radius minus a constant, B, times the partial charge on that atom. As a starting point in the evaluation of B, Pauling's method (A7) of dividing the internuclear distance in sodium fluoride between the two "ions" inversely proportionally to their effective nuclear charge was adopted in principle. However, a correction was made for the fact that the atoms are not truly ions, the fluorine atom having only about 9.8 electrons and the sodium atom 10.2. Values of B were determined from the radii of Na and F and the partial charges in NaF. These in turn were applied to determining the radius of sodium in the other halides, and the radius of fluorine in the other alkali metal fluorides. From these, subtraction from the experimental bond lengths gave radii of combined halogen and metal, which could be used for computing the value of B for each element. A general averaging procedure then provided the best values of B for all these elements. These values were used to obtain values of B for other major group elements. The results are summarized in Table 15:2.

The radii of the combined atoms and the bond lengths were then calculated for a number of inorganic halides. Results were quite satisfactory for the halides of elements the

Table 15:3

Apparent Limiting Radii of Combined 18–Shell Elements

	r_o	r_c		r_o	r_c		r_o	r_c
Zn	98.2	129.2	Ga	108.	125.6	Pb	135.0	148.0
Cd	113.5	149.3	In	127.5	145.5	As	111.2	119.4
Hg	120.5	150.0	Ge	106.5	122.3	Sb	131.4	138.9
Cu	90.1	133.1	Sn	127.7	142.0	Bi	143.9	147.0
Ag	107.1	153.3						

atoms of which have 8 electrons in the shell beneath the outermost, but not for those having 18 electrons in that level. It appears that these 18-shell elements exhibit a certain contraction beyond which they contract no further, depite increasing partial positive charge. If these elements are assigned a fixed average radius, the bond lengths are satisfactorily represented as the sum of this radius and the calculated radius of the appropriate halogen atom. Table 15:3 summarizes these limiting radii together with the corresponding nonpolar radii.

The calculated radii and bond lengths are given for the gaseous alkali metal halides in Table 15:4 and for the solid halides in Table 15:5. In general, the agreement between calculated and experimental bond lengths is very satisfactory, averaging within about one per cent. Data for the gaseous M2 halides are given in Table 15:6. Table 15:7 lists only the solid halides of Sr and Ba since for some unknown reason good values of B could not be obtained for Mg and Ca. Other gaseous halides having M8 structures beneath the valence shell are given in Table 15:8. Similar data for the 18-shell halides are given in Table 15:9.

In more complex molecules, sphericity of the atoms is probably not maintained and distortions must complicate the calculation of bond lengths. For all 135 bond lengths given in the tables the agreement between calculated and experimental bond lengths is within an average of 3 pm. There are 8 compounds with discrepancies greater than 10 pm, for reasons unknown. If these are eliminated from the averaging, the remaining 94 per cent show agreement averaging about 2 pm. This success must be regarded as supporting

Table 15:4

Calculated Radii and Bond Lengths in M1 Halide Gas Molecules

cmpd	charge	r_M	r_X	$r_M + r_X$ (pm)	R_o
LiF	0.753	46.4	107.3	153.7	156.4
NaF	0.797	80.2	109.5	189.7	192.6
KF	0.851	106.0	112.4	218.4	217.
RbF	0.920	112.3	115.9	228.2	227.
CsF	0.977	119.8	118.9	238.7	234.5
LiCl	0.666	56.5	145.2	201.7	202.1
NaCl	0.711	88.1	148.3	236.4	236.1
KCl	0.764	115.2	152.0	267.2	266.7
RbCl	0.833	122.1	156.7	278.8	279.
CsCl	0.890	130.1	160.6	290.7	290.6
LiBr	0.621	61.7	155.7	217.4	217.0
NaBr	0.666	92.3	158.7	251.0	250.2
KBr	0.718	120.0	162.2	282.2	282.1
RbBr	0.789	127.1	166.9	294.0	294.5
CsBr	0.846	135.3	170.7	306.0	307.2
LiI	0.540	71.1	169.5	240.6	239.
NaI	0.585	99.8	172.5	272.3	271.1
KI	0.638	128.6	176.0	304.6	304.8
RbI	0.711	135.9	180.9	316.8	317.7
CsI	0.764	144.9	184.5	329.4	332.

the validity of partial charge in a very satisfactory manner.

When all the quantitative evidence supporting the reality and accuracy of partial charge is carefully and open-mindedly considered, rejection of this valuable concept seems almost impossible to understand. It should not be at all surprising that partial charge has profound effects on the properties of compounds, as will be reviewed in the following pages.

Summary of
EVIDENCE SUPPORTING VALIDITY OF PARTIAL CHARGES

I. They have been proven INDISPENSABLE in the accurate CALCULATION OF ENERGIES of 15,000 bonds between more than 200 different pairs involving 38 major group elements in more than ONE THOUSAND molecular and nonmolecular compounds, both INORGANIC and ORGANIC.

II. In 191 examples, partial charges calculated from ELECTRONEGATIVITIES differ from those calculated INDEPENDENTLY FROM EXPERIMENTAL DATA ONLY by an average of only 0.007 charge unit.

III. By use of a simple linear relationship between partial charge and the radius of a combined atom, BOND LENGTHS for 136 bonds in gases and solids, involving 32 different elements, have been calculated within an average of 1.4 per cent of the experimental value.

IV. Bond energies based on these partial charges have been used successfully to calculate REORGANIZATIONAL ENERGIES of 27 organic and 154 inorganic radicals, and then 268 organic and 160 inorganic BOND DISSOCIATION ENERGIES.

V. PARTIAL CHARGES on COMBINED ATOMS have long been shown to be consistent with the PHYSICAL AND CHEMICAL PROPERTIES of compounds, and uniquely useful in INTERPRETING THEIR CHEMISTRY.

Table 15:5

Calculated Radii and Bond Lengths in Solid M1 Halides

cmpd	charge	r_M	r_X	$r_M + r_X$ (pm)	R_o
LiF	0.753	79.3	139.0	218.3	200.9
LiCl	0.666	85.6	176.4	262.0	257.
LiBr	0.621	88.8	186.7	275.5	275.
LiI	0.540	94.7	202.4	297.1	300.2
NaF	0.797	94.6	143.1	237.7	231.
NaCl	0.711	101.0	181.7	282.7	281.4
NaBr	0.666	104.3	191.9	296.2	298.0
NaI	0.585	110.4	208.2	319.6	323.1
KF	0.849	118.5	148.0	266.5	266.4
KCl	0.762	126.5	187.6	314.1	313.8
KBr	0.718	130.5	198.0	328.5	328.5
KI	0.637	137.9	214.8	352.7	352.5
RbF	0.918	123.7	154.5	278.2	282.
RbCl	0.831	132.5	195.5	328.0	326.7
RbBr	0.787	136.9	206.0	342.9	341.8
RbI	0.706	145.0	223.7	368.7	365.5
CsF	0.975	144.6	159.8	304.4	300.4
CsCl	0.889	152.6	202.3	354.9	356.6
CsBr	0.844	156.8	212.7	369.5	371.3
CsI	0.763	164.3	231.0	395.3	394.7

PARTIAL CHARGES IN POLYATOMIC IONS

It is useful to determine partial charges not only in molecules and nonmolecular solids, as described earlier, but also in polyatomic ions. Many years ago (B1) I attempted to devise a valid procedure for doing this. The method was to treat a polyatomic group, such as acetate, as though it were a neutral unit. For example, the electronegativity

Table 15:6

Calculated Radii and Bond Lengths in Gaseous M2 Halides

cmpd	charge on: M	X	r_M	r_X	$r_M + r_X$	R_o
BeF$_2$	0.592	-0.296	52.5	83.5	136.	140.
BeCl$_2$	0.464	-0.232	60.3	115.4	176.	177.
BeBr$_2$	0.400	-0.200	64.2	127.6	192.	191.
BeI$_2$	0.282	-0.141	71.4	142.7	214.	212.
MgF$_2$	0.788	-0.394	88.6	87.9	177.	177.
MgCl$_2$	0.656	-0.328	96.2	122.0	218.	218.
MgBr$_2$	0.588	-0.294	100.4	133.8	234.	234.
MgI$_2$	0.468	-0.234	108.0	149.0	257.	257.
CaF$_2$	0.972	-0.486	112.8	93.4	206.	210.
CaCl$_2$	0.836	-0.418	121.3	128.2	250.	251.
CaBr$_2$	0.766	-0.383	125.7	139.8	266.	263.
CaI$_2$	0.640	-0.320	133.7	154.7	288.	287.
SrF$_2$	1.108	-0.554	124.6	96.9	222.	220.
SrCl$_2$	0.968	-0.484	132.9	132.7	266.	267.
SrBr$_2$	0.898	-0.449	137.0	144.2	281.	282.
SrI$_2$	0.768	-0.384	144.7	159.0	304.	303.
BaF$_2$	1.156	-0.578	142.4	98.2	241.	232.
BaCl$_2$	1.016	-0.508	149.1	134.3	283.	282.
BaBr$_2$	0.944	-0.472	152.6	145.7	298.	299.
BaI$_2$	0.814	-0.407	158.8	160.6	319.	320.

of silver acetate is determined, and also of the acetate group.
The partial charge on silver is calculated in the usual manner.
The difference between the electronegativity of silver acetate
and acetate group is then divided by the partial charge on
silver, assuming a univalent anion, to learn how much the
electronegativity of the acetate ion would change in acquiring
a complete electron, and thus to determine the electronegativity of the ion. The reason for choosing the silver salt is that

Table 15:7

Calculated Radii and Bond Lengths in Solid M2 Halides

cmpd	d_M	$r_M(I)$	$r_M(II)$	d_X	r_X	$r_M(I)+r_X$	$r_M(II)+r_X$	R_o
SrF_2	1.108	142.1	141.9	-0.554	114.7	256.8	256.6	251.
$SrCl_2$	0.968	148.3	148.2	-0.484	153.8	302.1	302.0	302.
$SrBr_2$	0.898	151.4	151.3	-0.449	169.5	320.9	320.8	321.
SrI_2	0.768	157.2	157.2	-0.384	184.8	342.0	342.0	342.
BaF_2	1.156	156.5	154.2	-0.578	116.7	273.2	270.9	268.
$BaCl_2$	1.016	161.5	161.5	-0.508	156.5	318.0	318.0	318.
$BaBr_2$	0.944	164.1	165.0	-0.472	172.3	336.4	337.3	338.
BaI_2	0.814	168.8	171.1	-0.407	187.9	356.7	359.0	359.

in salts of active metals, electronegativity equalization cannot occur, because the metal atom would have to lose more than its normal valence electrons to become as electronegative as the anion. For example, a sulfate ion is more electronegative than a sodium ion, which means that the partial charge on sodium if equalization were to occur in Na_2SO_4 would be greater than one. This is impossible. There seems no alternative in such cases but to consider the salt completely ionic, but with the anion probably distorted or polarized by the adjacent cations.

This method seemed acceptable for obtaining relative values which could be usefully compared, but too arbitrary to justify confidence in the reality of the values. It is therefore interesting that newly developed, much simpler method, reported here for the first time, gives almost identical results. The new procedure involves the recognition that there can be only one value of electronegativity which corresponds to partial charges which add up exactly to the charge on the ion. It is therefore necessary only to choose arbitrarily one lower and one higher electronegativity for the polyatomic ion and calculate the partial charges and thus the total charge in both. Linear interpolation between the two total charges to the actual charge on the ion allows determination of the appropriate electronegativity of the cation or anion, since

Table 15:8

Calculated Radii and Bond Lengths in Other 8-Shell Gaseous Halides

cmpd	charge on:		r_M	r_X	$r_M + r_X$	R_o
	M	X				
HF	0.248	-0.248	7.9	81.0	88.9	91.8
HCl	0.162	-0.162	16.3	110.5	126.8	127.4
HBr	0.117	-0.117	20.6	122.0	142.6	140.8
HI	0.036	-0.036	28.5	135.7	164.2	160.8
BF_3	0.504	-0.168	55.7	76.8	132.5	129.5
BCl_3	0.357	-0.119	63.4	107.6	171.0	174.
BBr_3	0.285	-0.095	67.2	120.5	187.7	187.
BI_3	0.156	-0.052	74.0	136.8	210.8	203.
AlF_3	0.729	-0.243	84.6	80.7	165.3	163.
$AlCl_3$	0.576	-0.192	93.3	112.6	205.9	206.
$AlBr_3$	0.507	-0.169	97.1	125.5	222.7	227.
AlI_3	0.369	-0.123	105.0	141.4	246.4	244.
SiF_4	0.600	-0.150	82.5	75.9	158.4	156.1
$SiCl_4$	0.440	-0.110	91.7	107.0	198.7	201.9
$SiBr_4$	0.360	-0.090	96.3	120.2	216.5	216.
SiI_4	0.216	-0.054	104.5	136.9	241.4	241.
PF_3	0.417	-0.139	91.9	75.9	167.8	157.
PCl_3	0.276	-0.092	98.2	105.7	203.9	204.
PBr_3	0.207	-0.069	101.3	118.7	220.0	220.
PI_3	0.078	-0.026	107.2	135.0	242.2	247.

the charge on the polyatomic ion is linear with its electro-negativity.

For example, if the electronegativity of sulfate ion were 2, the partial charges on S, -0.354, and O, -0.551, would total -2.558 for the group. If the electronegativity of sulfate ion were 3, the partial charges of S 0.016 and O -0.218 would total -0.856. The linear equation derived from these two

Table 15:9

**Calculated Radii and Bond Lengths in Gaseous Halides
of 18-Shell Elements**

cmpd	charge on:		r_M	r_X	$r_M + r_X$	R_o
	M	X				
CuF	0.366	−0.366	90.1	87.1	177.	175.
CuCl	0.279	−0.279	90.1	118.6	209.	205.
CuBr	0.235	−0.235	90.1	129.9	220.	
CuI	0.153	−0.153	90.1	143.6	234.	240.
AgF	0.413	−0.413	107.1	89.6	197.	198.
AgCl	0.327	−0.327	107.1	121.9	229.	228.
AgBr	0.282	−0.282	107.1	133.0	240.	239.
AgI	0.201	−0.201	107.1	146.8	254.	254.
ZnF_2	0.452	−0.226	98.2	79.9	178.	181.
$ZnCl_2$	0.328	−0.164	98.2	110.7	209.	205.
$ZnBr_2$	0.266	−0.133	98.2	123.1	221.	221.
ZnI_2	0.152	−0.076	98.2	138.4	237.	238.
CdF_2	0.544	−0.272	113.	82.2	195.	
$CdCl_2$	0.406	−0.203	113.	113.4	226.	224.
$CdBr_2$	0.342	−0.171	113.	125.6	239.	239.
CdI_2	0.228	−0.114	113.	140.9	254.	256.
HgF_2	0.462	−0.231	120.5	80.1	201.	196.
$HgCl_2$	0.336	−0.168	120.5	111.0	232.	234.
$HgBr_2$	0.274	−0.137	120.5	123.4	244.	244.
HgI_2	0.160	−0.080	120.5	138.7	260.	261.
GaF_3	0.456	−0.152	107.7	76.0	184.3	188.
$GaCl_3$	0.312	−0.104	107.7	106.6	214.2	209.
$GaBr_3$	0.240	−0.080	107.7	119.5	227.2	228.
GaI_3	0.114	−0.038	107.7	135.8	243.5	244.
InF_3	0.516	−0.172	127.5	77.0	205.2	189.
$InCl_3$	0.372	−0.124	127.5	107.9	235.4	234.
$InBr_3$	0.297	−0.099	127.5	120.7	248.2	247.
InI_3	0.168	−0.056	127.5	137.1	264.6	267.

Table 15:9 cont.

| cmpd | charge on: | | r_M | r_X | $r_M + r_X$ | R_o |
	M	X				
GeF_4	0.416	-0.104	106.4	73.5	179.9	168.
$GeCl_4$	0.264	-0.066	106.4	103.9	210.3	208.
$GeBr_4$	0.188	-0.047	106.4	117.3	223.8	229.7
GeI_4	0.052	-0.013	106.4	134.2	240.6	249.
SnF_4	0.535	-0.134	127.7	75.1	202.8	186.
$SnCl_4$	0.352	-0.088	127.7	105.4	233.1	233.
$SnBr_4$	0.272	-0.068	127.7	118.7	246.4	246.
SnI_4	0.160	-0.040	127.7	135.9	263.6	264.
PbF_4	0.536	-0.134	135.0	75.1	210.1	208.
$PbCl_4$	0.380	-0.095	135.0	105.9	240.9	243.
AsF_3	0.321	-0.107	111.2	73.7	184.9	171.
$AsCl_3$	0.183	-0.061	111.2	103.6	214.8	216.
$AsBr_3$	0.114	-0.038	111.2	116.7	227.9	233.
AsI_3	0.012	-0.004	111.2	133.0	244.2	252.
SbF_3	0.438	-0.146	131.4	75.7	207.1	190.
$SbCl_3$	0.294	-0.098	131.4	106.1	237.5	233.
$SbBr_3$	0.225	-0.075	131.4	119.1	250.5	250.
SbI_3	0.096	-0.032	131.4	135.4	266.8	272.
BiF_3	0.480	-0.160	143.9	76.4	220.3	196.
$BiCl_3$	0.333	-0.111	143.9	107.0	250.9	248.
$BiBr_3$	0.262	-0.087	143.9	120.0	263.9	263.
BiI_3	0.132	-0.044	143.9	136.2	280.1	284.

points is: S = 0.588 times ionic charge + 3.503. For the -2 charge, S = 2.327. The partial charges corresponding to this electronegativity are S -0.233 and O -0.442, which summed for SO_4 gives -2.002 for the charge on sulfate ion. In this manner, the data of Table 15:10 were obtained, and an unlimited number of additional values could be derived.

Table 15:10

Electronegativities and Partial Charges in Polyatomic Ions

ion	S	partial charges			sum over ion
CH_3^+	3.266	0.200	0.267		1.000
CH_3^-	1.993	-0.289	-0.237		-1.000
NH_4^+	3.219	0.009	0.248		1.001
NH_2^-	1.908	-0.458	-0.271		-1.000
H_3O^+	3.483	0.352	-0.057		0.999
OH^-	1.705	-0.649	-0.351		-1.000
$CO_3^=$	1.956	-0.304	-0.566		-2.002
HCO_3^-	2.660	0.027	-0.033	-0.331	-1.000
$C_2O_4^=$	2.369	-0.145	-0.428		-2.003
CN^-	1.612	-0.436	-0.564		-1.000
NCS^-	2.061	-0.404	-0.263	-0.332	-0.999
NO_2^-	2.516	-0.242	-0.379		-1.000
NO_3^-	2.796	-0.142	-0.286		-1.000
$PO_4^{\equiv}$	1.660	-0.343	-0.664		-2.999
$HPO_4^=$	2.303	-0.114	-0.084	-0.450	-1.998
$H_2PO_4^-$	2.744	0.060	0.092	-0.303	-1.000
$SO_3^=$	2.005	-0.353	-0.549		-2.000
HSO_3^-	2.703	0.044	-0.094	-0.317	-1.001
$SO_4^=$	2.327	-0.233	-0.442		-2.002
HSO_4^-	2.854	0.104	-0.038	-0.267	-1.002
FHF^-	2.492	-0.480	-0.040		-1.001
ClO^-	2.082	-0.524	-0.476		-1.000
ClO_3^-	2.862	-0.209	-0.264		-1.001
ClO_4^-	3.020	-0.155	-0.211		-1.000
BF_4^-	2.980	0.298	-0.325		-1.002
CH_3COO^-	2.525	-0.085	-0.027	-0.376	-1.002

It seems likely that equalization of electronegativity does not necessarily occur in the formation of coordinate covalent bonds, as discussed in Chapter 14. Therefore although the evaluation of partial charge in polyatomic ions such as sulfate or ammonium is satisfactory, it may not be so in complex ions involving molecular ligands such as H_2O or NH_3. For example, assuming electronegativity equalization in the tetraaquo and tetrammine zinc ions, the electronegativities are determined as, $Zn(H_2O)_4^{++}$ 3.254, and $Zn(NH_3)_4^{++}$ 3.002. Data for bond energy calculations are not available, but the ionic blending coefficient t_i is 0.287 for Zn–O and 0.201 for Zn–N. These suggest a much stronger bond to oxygen, which appears to be opposite to fact. It may be concluded tentatively at least that electronegativity equalization does not occur in such complex ions.

PHYSICAL AND CHEMICAL RELATIONSHIPS TO PARTIAL CHARGE

The acquisition of partial charge influences chemical properties, and particularly those of oxidation–reduction and acid–base reactions, in a reasonable manner. Consider, for example, an atom of a highly electronegative element. In keeping with its high electronegativity, it has its outermost principal quantum level more than half filled to its quota of 8 electrons. The effective nuclear charge is relatively high and can be sensed within one or a few outer vacancies. The atom therefore has the capacity to act as oxidizing agent toward less electronegative atoms. Among other things, this means it has the ability on combining with hydrogen to give it a substantial positive charge with corresponding acidity resulting. However, as this atom succeeds in acquiring the electron(s) for which it is initially hungry, its electron-hunger must diminish. Thus it gradually loses its oxidizing power and undergoes a general change in character. By the time it has become quite negative, it is no longer an oxidizing agent, and in fact, if it was not too electronegative in the first place, may have acquired some reducing power. The outer electron pairs were originally held too tightly to be readily available for electron pair donation, but become more available with increasing partial negative charge. They may have appreciable donating power by the time the atom becomes quite negative. In other words, an atom that is initially oxidizing and perhaps acidic tends, with increasing negative charge, to become less and less so and a better base and electron pair donor. It loses its oxidizing power, and if not too electro-- negative initially, becomes reducing.

In contrast, consider an atom of an active metal. Its metallic nature is derived from the surplus of outer vacancies, so these must be kept in mind for their potential chemistry. In keeping with its relatively few outer electrons, the atom must be low in electronegativity and capable of reacting with nonmetal atoms only by providing electrons to them. In other words, the atom has reducing power, and certainly no oxidizing ability of any consequence. In its general nature it tends to be basic, and not at all acidic. However, as it begins to lose its valence electrons, the situation changes. The remaining electronic cloud tends to be pulled in more tightly around the nucleus as the inter-electronic repulsions are reduced and the effective nuclear charge increases. Thus the electronegativity increases, and all reducing power is lost, oxidizing power tending to take its place. With outer vacant orbitals and appreciable positive charge, the atom becomes a potential electron pair acceptor, and in this sense acidic.

It should be clear that when atoms combine to form compounds in which their previously neutral condition is altered by the acquisition of partial positive or negative charge, the properties of these atoms must change conspicuously from those of the free atoms. A notable example of this is of course sodium chloride, often cited to beginning students as exemplifying the magical changes that occur when atoms form compounds. All polar covalence, however, is of this same nature, even though not always so obviously or spectacu- larly so.

These are some general effects of partial charge on the chemical properties. Physical properties change as well, although not necessarily so much because of the partial charges as because of the characteristics that go along with the partial charge. For example, all metal atoms, when they have com- bined with nonmetal atoms, still retain vacant outer orbitals in the simplest molecules that might form. All nonmetal atoms except those of Group M4, the carbon family, have lone pair outer electrons in addition to their bonding electrons, and these are still uninvolved in the simplest molecules.

The individual molecules of metal-nonmetal compound can therefore not only attract one another through the partial positive and negative charges, but also by coordination, the electron pairs of the nonmetal being attracted into the receptive vacant orbitals of the metal. The result is the formation of nonmolecular solid, which means relatively high melting point and low volatility. But as the bond strength decreases owing to reduced polarity in such solids, the temperature of melting is reduced and the volatility increased. There is relatively little opportunity for nonmetal-nonmetal molecules to condense unless they can polymerize, so these compounds, if solid, are usually molecular solids having to diminish or overcome only van der Waals forces in order to melt or vaporize. Consequently there is a definite relationship, although not necessarily directly cause-and-effect, between physical properties, such as melting and boiling points, and partial charge.

The most obvious property to be influenced by the charge distribution is of course the dipole moment. One cannot expect to calculate dipole moment accurately from partial charges, because such moments are very susceptible to relatively slight shifts in the electronic clouds. For example, dipole moments measured for HF have led to estimates of ionicity that are far too high. Although usually dipoles are interpreted as resulting from partial charges at the nuclei, actually it is the distance between the total positive charge center and the center of the entire electronic cloud that determines the dipole moment. In HF there are 10 electrons and a total of 10 positive charges, and the distance between the centers of these is far less than the bond length.

FUNCTIONALITY

From organic chemistry the condition of functional groups as the more negative components of the molecules is recognized. The same holds for inorganic chemistry. That is, the chemical characteristics of inorganic compounds with positive and negative components are largely, although of course not exclusively, a function of the condition and nature of the negative component, which may be regarded as the functional group. A group of compounds of chlorine, for example, in all of which the chlorine bears high negative charge, resemble one another much more than a group of compounds of any particular metal, all with different negative components. For this reason, although there are enough advantages to studying the chemistry of the elements by periodic groups to make this approach practically universal, there are, in my opinion, more advantages to considering the chemistry of the nonmetals over the entire periodic system. This was the approach used for my earlier textbooks of inorganic chemistry (B1, B3), and will be continued here. The following chapter reviews the principal features of such a study, which seems to me pedagogically and logically superior to the conventional one. Specific examples of the effects of partial charge in specific classes of compounds will be discussed. Finally, methods of classroom presentation with the aid of models will be reviewed, for they can be extremely helpful in explaining chemistry.

Chapter 16

PERIODICITY AND PARTIAL CHARGE

Debate has continued over many years regarding the relative roles of principles and practice in the teaching of general and inorganic chemistry. Those who would relegate "descriptive" chemistry to a minor role commonly argue that a recital of chemical facts is boring and soon forgotten. Those who favor more descriptive chemistry insist that it is the practical applications of chemistry which give it such universal importance. Those who try to view both the forest and the trees recognize the importance of principles as establishing a framework upon which to build the structure of descriptive facts, and the importance of facts to give reality to the principles. The principal cause of the situation that evokes this debate, in my opinion, is the failure appropriately to **integrate theory with practice.** I think such integration can readily be accomplished by **application of the simple concepts of electronegativity, partial charge, and bond energies** described in this book. The ingredients have been available for many years. All that is needed is a willingness to try them. There is no excuse for presenting chemistry as a hodge-podge of miscellany, when it can be a beautifully organized, self-consistent and logical science, intelligible by ordinary mortals.

BINARY OXIDES

Being surpassed in electronegativity only by fluorine, oxygen acquires partial negative charge in all combinations except with fluorine. Where the oxygen is joined to other also highly electronegative elements (N, Cl, Br, S, Se), the partial negative charge it is able to acquire can only be small. The oxide (SO_3, Cl_2O_7) will therefore retain considerable oxidizing power. With water, the oxide will act as an acid, forming a hydroxy compound in which the hydrogen is quite positive and acidic. Being composed of molecules having no opportunity to condense further, or at least not very stably, and having relatively slightly polar bonds, the oxide will be volatile. Then, as the opportunity for condensation increases, the bonds becoming more polar and vacant orbitals becoming available, the oxide tends to become less volatile and lose its oxidizing power. The acidity is reduced, and presently the oxide is amphoteric. With still higher partial negative charge on the oxygen, the oxide tends to become less acidic and more basic in its amphoterism, less volatile, and essentially nonoxidizing. Finally, with high partial negative charge on oxygen, the compounds are relatively high melting, totally nonacidic, highly basic, and although nonoxidizing, not really reducing, although able to form strong coordination bonds as an electron pair donor. Partial charge thus provides the explanation of the old rule that nonmetal oxides tend to be acidic and metal oxides basic, and it also explains the diminishing basicity and increasing acidity of oxides as the positive oxidation state of the other element increases, creating competition among oxygen atoms that prevents any one of them from becoming very negative. The absurdity of the practice of estimating the nature of bonds merely by noting the difference in electronegativity between the two elements is well illustrated by the example of the H–O bond in NaOH and $HOClO_3$. This practice fails to differentiate between the two.

Partial charges on oxygen and their consequences are represented in Figure 16:1. The periodicity of partial charge is of course obvious in the figure, as also the periodicity of oxide properties. It should not be surprising that oxides of opposite nature would react with one another, forming complex oxysalts. For example, the oxygen in CaO has high negative charge (-0.60), and is able to donate an electron pair to the positive sulfur (0.19) in SO_3 to form $CaSO_4$. It should also be apparent that an oxide with intermediate partial negative charge on oxygen might be expected to react either way. It may act as an acidic oxide toward an oxide having higher partial negative charge or as a basic oxide toward an oxide having a lower partial negative charge. For example, ZnO with O -0.27 can form sodium zincate, Na_2ZnO_2 with Na_2O, d_O -0.83, or it can form $Zn(NO_3)_2$ by reaction with N_2O_5, d_O -0.05.

Water, by virtue of its liquid condition and excellent solvent properties, has often been described more for its action as solvent than for its typical binary oxide behavior. From the latter viewpoint, water is simply an amphoteric oxide of hydrogen, d_O -0.25. It can unite with an oxide having higher negative charge on oxygen, such as BaO, d_O -0.70, to form the basic hydroxide, $Ba(OH)_2$, or with an oxide having

Figure 16:1

Partial Charge on Oxygen in Oxides
(in highest oxidation state)

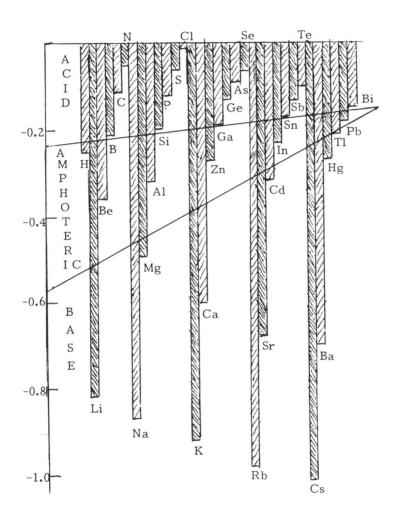

lower negative charge, such as Cl_2O_7, δ_O -0.01, to form an acidic hydroxy compound, perchloric acid, $HOClO_3$. In this sense it is not significantly different from the amphoteric zinc oxide. However, traditionally, the complex oxysalts of hydrogen have been defined as bases in such compounds as

NaOH and acids in such compounds as $(HO)_2SO_2$, instead of hydrogen salts, and so the resemblance between water and other amphoteric oxides is often overlooked.

The binary chemistry of oxygen, when studied from the viewpoint of partial charges, becomes far more meaningful than when merely viewed as a collection of facts to be memorized. Students can become much more interested in learning when what they learn makes sense and fits a logical, understandable pattern.

HALIDES

M7 elements, the halogens, offer the most abundant data for comparisons in the entire periodic system. Binary compounds are formed between the halogens and nearly all of the other elements. Meaningful comparisons can be made among the halides, revealing certain systematic similarities and certain differences. The similarities arise from the similarity among the halogens, all consisting of atoms having the outermost principal quantum level nearly filled, with seven electrons, allowing one vacancy and resulting in the highest electronegativity in each period. The differences among binary halides result from the increasing radius and decreasing electronegativity from fluorine to iodine. Any similarities are much less conspicuous among halogen compounds in which they exhibit positive oxidation states, where each halogen is quite individualistic. But especially in the halides, the comparisons are greatly aided by knowledge of the partial charges on halogen atom.

The combination of **decreasing electronegativity** and **increasing radius** down the halogen group results in stronger bonds in fluorides than in chlorides, which in turn have stronger bonds than in bromides, which in turn have stronger bonds than in iodides. These stronger bonds correspond to larger partial negative charges on halogen. They also correspond to generally higher melting, less volatile fluorides compared to chlorides, bromides, and then iodides. The reducing power of a combined atom is ordinarily enhanced by partial negative charge, but atoms which were initially very high in electronegativity are not expected to release electrons easily, even when quite negative. Iodine in its iodides is much more readily a reducing agent than the other halides in which the negative charge on halogen is higher, and fluoride is not reducing at all. All highly negative halogen atoms, however, have the capacity to serve as electron pair donors in the formation of complexes. One more point worth noting is that the smaller size of fluorine permits it to form higher halides than the other halogens, for example, as in SF_6 for which there is no other halogen counterpart.

Having duly noted these fundamental differences among the different kinds of halides, we may now consider the similarities. The periodic trends within fluorides, chlorides, bromides, and iodides, are all very much alike, and not unlike those of oxides. Figures 16:2 and 16:3 show partial charges on halogen in binary fluorides and chlorides. Halides in which the halogen has been able to acquire only small partial negative charge are volatile and retain some of their halogenating ability. With increasing partial negative charge on halogen, a trend is observed toward less oxidizing, less volatile, to nonoxidizing and high melting nonmolecular solids. A typical reaction of halides is hydrolysis. This follows the nature of the hydroxides which, along with hydrogen halide, are a product. If the hydroxide is exclusively acidic, then there is no chance of reversible hydrolysis and it proceeds to completion. If the hydroxide is amphoteric there can be reversible hydrolysis, and if the hydroxide ionizes exclusively as a base, the hydrolysis is less extensive if the basicity is weak, and nonexistent if the base is strong.

Again, it seems far more usefully instructive to point out these trends and their basic causes than to provide merely a recital of properties of individual halides without questioning, much less explaining, the origin of these properties. The functional nature of the halogen in halides is illustrated by the following: If the halogen atom has a very high partial negative charge in any combination with another kind of element, no matter what that element is, it may confidently be predicted that the compound will be a high melting, nonvolatile nonmolecular solid having no oxidizing power and able to form complexes by acting as electron pair donor. It will not be susceptible to hydrolysis. On the other hand, any halide in which the halogen has not acquired very much negative charge must be molecular, volatile, oxidizing, and very susceptible to hydrolysis.

Figure 16:2

Partial Charge on F in Binary Fluorides

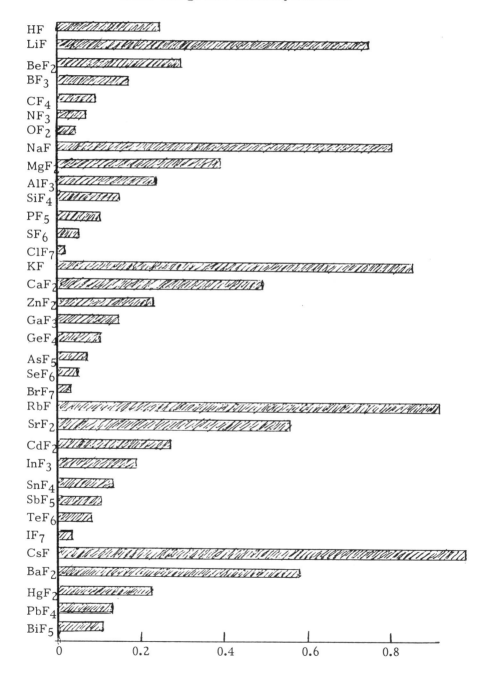

Figure 16:3

Partial Charge on Cl in Binary Chlorides

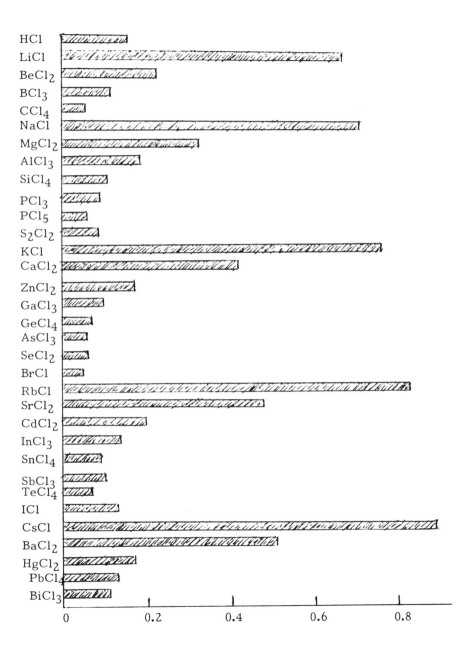

BINARY COMPOUNDS OF HYDROGEN

By virtue of its electronegativity being intermediate among the values of the other major group elements, hydrogen has an extensive chemistry in both directions. That is, there are a number of other elements having lower electronegativity with which hydrogen forms compounds containing partially negative hydrogen, properly called hydrides. There are also a number of elements having higher electronegativity with which hydrogen can combine acquiring partially positive charge, sometimes improperly called "hydrides" but better referred to merely as binary compounds of positive hydrogen. These two types of hydrogen compound, hydridic and protonic, are of chemically opposite nature, and therefore react with one another.

Consider first the properties of hydrogen itself. It has the property of an oxidizing agent in combination with all the elements which are less electronegative. It has the property of a reducing agent in combination with all the elements which are more electronegative. However, it can be neither donor nor acceptor. This is hydrogen with zero charge initially. Now, if it gradually gains partial negative charge, the reducing quality must be expected to increase, and the oxidizing quality to decrease, until finally highly negative hydrogen is strongly reducing and not oxidizing at all, and highly positive hydrogen is oxidizing and not reducing at all. Furthermore, highly negative hydrogen has the capacity in combination with highly positive metal atoms to form stable nonmolecular solids. Highly positive hydrogen must be in combination with highly electronegative elements which have no outer vacancies permitting further condensation. Hence these compounds are molecular and volatile. Highly negative hydrogen is basic and able to serve as electron pair donor in the formation of complex hydrides. Highly positive hydrogen is acidic and able to act as electron pair acceptor as a proton in the formation of complexes. However, when highly negative and highly positive hydrogen come together, they form molecular hydrogen, H_2, which is liberated. Figure 16:4 shows the partial charge on H in binary hydrogen compounds.

An additional interesting fact about hydrogen is that it is the only element whose atoms use all their electrons in

Figure 16:4

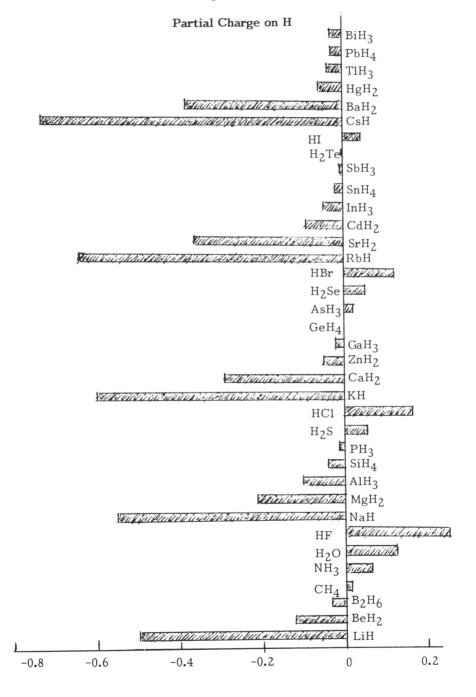

bonding. When the hydrogen electron is involved in a covalent bond which imparts a partial positive charge, it must be concentrated within the bonding region between the nuclei, leaving no appreciable electronic cloud to protect its nucleus. Hence the protonic end of the hydrogen atom, opposite the bond, is susceptible to attraction by and for an electron pair on some other atom of another molecule. Thus the hydrogen can form a bridge between its own and a neighboring molecule. The bridge is usually linear but unsymmetrical, the proton remaining much closer to its original bond. Such bridging is conventionally called "hydrogen bonding." For reasons to become apparent, it should more sensibly be termed "protonic bridging," to indicate the importance of the partial positive charge on the hydrogen and to distinguish it from other types of hydrogen bridging. In a typical protonic bridge, the hydrogen bears a significant partial positive charge, and the atom to which it forms the bridge is small, provides at least one outermost electron pair, and bears a partial negative charge.

The other atom must be small because on a larger atom the electron pair appears to be more spread out and less able to attract the positive hydrogen. The most common protonic bridges by far are therefore to N, O, or F. The other atom must also be substantially negative because otherwise, despite the existence of an outer electron pair, it cannot make the electron pair available to the hydrogen. For example, in the hydronium ion, H_3O^+, the partial charge on oxygen, only -0.06, is insufficient to form a protonic bridge, and the hydronium ion can become "trihydrated" by formation of protonic bridges using the three very positive hydrogen atoms, producing the species, $(H(H_2O)_4^+$ wherein three of the water molecules are held much less tightly than the fourth. Other hydrates are also possible.

Neutral or only slightly positive hydrogen does not form protonic bridges. For example, methane does not form such bridges, but chloroform, $CHCl_3$, in which the hydrogen bears higher partial positive charge, does.

Partially negative hydrogen can also form bridges, but these are of quite a different kind. They consist of three-center bonds in which the negative hydrogen shares a pair of electrons with two other atoms, each of which provides a vacant orbital and a partial positive charge. The best known example is the bridging in diborane, by which two BH_3 groups are caused to form B_2H_6. Actually two BH_2 groups are held together by two bridges, which because of the negative hydrogen may be termed "hydridic bridges." Whereas an average **protonic bridge** features a **protonic** hydrogen holding together two **negative** atoms each providing a **pair of electrons,** a **hydridic bridge** consists of a **hydridic** hydrogen with **two electrons** holding together two **positive** atoms each providing a **vacant orbital.** The usual **protonic bridge** is **linear,** with the hydrogen much **closer** to its own molecule than to the other molecule, and the bond energy is about 3-10 kcal (12-42 kJ) per mole. The usual **hydridic bridge** is **nonlinear,** with the hydrogen placed **symmetrically** with respect to the two bridged atoms, and it is about 3-6 times stronger than a protonic bridge, being, in effect, a half-bond.

Partial charge has permitted recognition of the different nature of these two types of bridging by hydrogen, but skepticism about partial charge, evidently, has prevented acceptance. Authors still use the term "hydrogen bond" quite indiscriminately. Furthermore, they still classify binary compounds of hydrogen as "ionic" and "covalent" or "saltlike (saline)" and "volatile", with bewildering disregard of a much more logical and meaningful approach.

Protonic bridges themselves exhibit a variety of types, which help to perpetuate the confusion. In some instances, the proton appears halfway between the two bridged atoms. Indeed, it is not always certain that such bridging is really protonic. For example, in the bifluoride ion, $F-H-F^-$, the hydrogen is midway between the two fluorine atoms in a linear ion. Assuming electronegativity equalization, the partial charge on hydrogen is not positive, but -0.040. In other words, far from having more than its capacity of 2 electrons around it, by forming two bonds, it averages only 1.040 electron.

Figure 16:4 summarizes the periodicity of partial charge on hydrogen and provides a sensible and systematic basis for discussion and describing the properties of binary compounds of hydrogen.

It is interesting to note that in the combination of hydrogen with elements less electronegative, it quite resembles the other nonmetals, oxygen and the halogens, in that highly negative hydrogen exists in nonmolecular solids, and slightly negative hydrogen in volatile compounds, and that highly negative hydrogen can act as electron pair donor but slightly negative hydrogen cannot. In the formation of complex hydrides, it is always the more negative hydrogen that acts as electron donor, and the atom combined with the less negative hydrogen that acts as electron pair acceptor. For example, AlH_3 acts as donor to BH_3, forming $Al(BH_4)_3$, in which three BH_2 groups are attached to Al by double hydridic bridges, but BH_3 does not donate to AlH_3 to form $B(AlH_4)_3$. The concept of partial charge and its importance in the formation of complex hydrides explains very well the failure of attempts to prepare complex hydrides of carbon and silicon, such as carbon borohydride, once sought as a possible high energy fuel.

IONIZATION OF WATER

One of the most rejected and useful applications of partial charge in hydrogen compounds is to an understanding of water and its ions. The partial charges on hydrogen and oxygen are as follows: H_3O^+ 0.352, -0.057; H_2O 0.124, -0.248; OH^- -0.351, -0.649. Writing the equation for the autoionization of water does not even suggest these differences. Water itself is amphoteric, but as the hydrogen becomes more positive and the oxygen less negative, the basic properties are lost and the acidic properties enhanced. Hydronium ion is not an electron pair donor at'all but it is strongly acidic, and an oxidizing agent toward good reducing agents. As the oxygen and hydrogen of water become more negative, the acidic properties are diminished and the basic property enhanced. In hydroxide ion, the hydrogen is highly negative and incapable of any type of acidic or protonic reaction under any ordinary conditions. The oxygen with its very negative charge becomes a much better electron pair donor than water.

When hydronium ion and hydroxide ion come together, a competition exists between a molecule of water, which already holds an extra proton, and a hydroxide ion, to determine which can **attract the proton more strongly.** Clearly this will be the oxygen with **higher negative charge.** Oxygen in OH^- with partial charge -0.65 is a much more effective donor to the proton than oxygen in H_2O with partial charge -0.25.

Therefore the hydroxide oxygen acquires the proton even though the water had it first. The process of **neutralization** thus becomes very clear and logical and quite to be expected.

What is not so readily apparent is why the ionization occurs in the first place. It must be remembered that protonic bridges in liquid water are continually forming and breaking, and it must be imagined that now and then a proton might accidentally find itself on the wrong side of the bridge. This would leave a hydronium and a hydroxide ion in close proximity, and one would expect the proton to jump back instantly to its original oxygen. Most of the time it probably does. However, a hydronium ion is simply a water molecule with an extra proton, and the proton can as readily sit on one water molecule as another. At the moment of formation of the hydronium ion, one of the other protons of this water molecule can easily transfer to an adjacent water molecule. Protons can travel easily from one water molecule to another, so in effect, the hydronium ion can migrate instantly from the location where it was formed. Similarly, a water molecule is nothing but a hydroxide ion with another proton added. The proton can sit with equal advantage on any hydroxide ion, so the instant a hydroxide ion is formed, it can attract a proton from an adjacent water molecule, and in effect, migrate from the place of formation. Protons can travel easily from one hydroxide ion to another, so without actual migration of the oxygen, the hydroxide ion can move rapidly through the liquid. Thus, although it is a little difficult to understand exactly how the transfer of a proton from one water molecule to another, which is the equivalent of moving a proton from a seat on a hydroxide ion to a seat on a water molecule, can occur, once it occurs, the immediate separation of the ions to prevent the reverse is readily imaginable. Not until the concentrations of hydronium and hydroxide ion reach 10^{-7} M does the rate of recombination reach the rate of formation, and is equilibrium attained.

STRENGTHS OF OXYACIDS

Partial charge is also useful in helping an understanding of acid strengths, particularly strengths of hydroxyacids. In an aqueous solution, the strength of an acid depends on a competition between the water molecules, which are normally present in great majority, and the anions of the acid. Both seek the proton of the acid. If the anion is a strong enough electron pair donor to the proton to resist the removal of the proton by water, the acid is weak. Most of it in solution is molecular and very little hydronium ion is formed. If the anion is not a strong enough electron pair donor to the proton, water molecules will attract the proton away, forming hydronium and leaving anion. Since the basicity of water is constant, the explanation of different acid strengths must lie in the nature of the anion formed from the acid. There are several factors that may influence the ability of the anion to donate to the proton, one of the most important of which is the partial charge on the donor oxygen. A good but imperfect correlation can be observed between the pK values for the acid and the partial charge on oxygen in its anion, as shown in Figure 16:5. As expected, the higher the partial negative charge on the oxygen, the weaker the acid.

The relatively small size of the fluoride ion allows it to attract a proton, strongly, which contributes to making HF a weak acid in water. HI is the strongest acid in this group, it being easiest to remove a proton from an electron pair on the relatively large iodide ion.

It should be clear from the equalization of electronegativity that additional nonmetal atoms attaching to the central atom will reduce the negative charge on all the nonmetal atoms and therefore weaken the basic properties of the anion, thus strengthening the acid. This accounts for such phenomena as the generally observed decrease of basicity and increase in acidity with higher positive oxidation state (actually higher partial positive charge) of the central atom. The most familiar example is the series of increasing acid strengths in $HOCl$, $HOClO$, $HOClO_2$, and $HOClO_3$. It also provides understanding of the hydrolysis of anions, those of weak acids hydrolyzing most extensively.

The artificiality of the system of oxidation numbers is also clearly shown by partial charge calculations. For example, the charge on phosphorus in PF_3 is 0.42, compared to only 0.31 in PCl_5, although the latter is considered to be in a higher oxidation state.

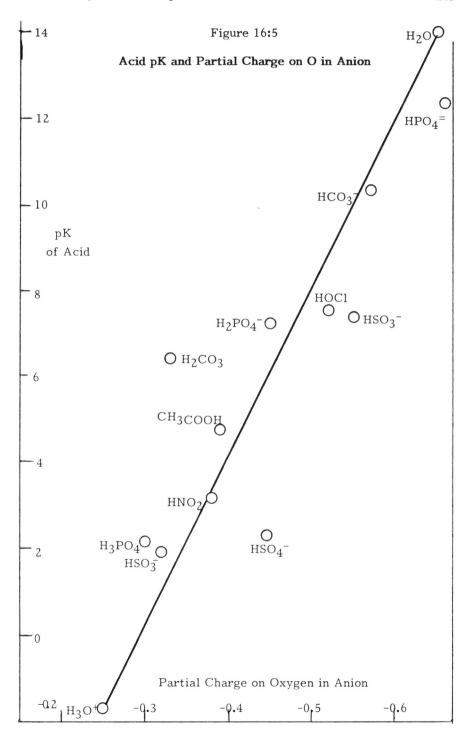

Figure 16:5

Acid pK and Partial Charge on O in Anion

VISUALIZATION WITH THE HELP OF MODELS

Purposes of Models

There are two principal types of physical model which can be very helpful in the classroom or independent study. One type attempts to reproduce, but on a convenient scale, something too small to be easily seen or too large to bring into the room for demonstration, for example, a flea or the solar system. In the case of machinery it is desirable to build not only as exact a model as possible, but also one that operates, a working model. The other type of physical model is one which makes no serious attempt to be identical with the real object, except for scale, but which represents the significant features in such a way as to allow better understanding through visualization. Such are the atomic and molecular models devised more than twenty years ago to which I now call renewed attention.

It is not within the power of mortal man to know what he will discover or in what order, although the orderly presentation of data in a scientific paper may create the illusion that the experimentation was equally well thought out and that all results were produced in a perfectly logical sequence. It would have been ideal if after evaluating electronegativity and partial charge, I had immediately understood how these data could be employed toward a better understanding of bond energies. Unfortunately, this process took more than a decade. Meanwhile, I devised models to represent electronegativity and partial charge and introduced them to chemistry teachers through lectures, films, and writings (B2, B19-21, B36-38) **without** the very important support of demonstrating their applicability to the quantitative calculation of the energies of polar covalent bonds. Although great interest was shown initially, the general enthusiasm appears to have become subdued by the skepticism of experts (and nonexperts) regarding the basic concepts. There seems to be a widely accepted rule of science that if an explanation can be understood, it is far too simple. Models that support and enhance such explanation are therefore apparently unacceptable.

My own personal belief is that the ultimate in understanding is forever inaccessible to man. The best we can hope for is to acquire **the illusion of understanding.** This can serve us well as long as (a) we are fully aware of **its limitations** and **our own,** and (b) it provides us with a **practical view that is helpful.** Such an illusion can exist in all degrees of sophistication, each appropriate for its purpose. Contrary to common belief, however, ignorance gains no special merit for existing at a high intellectual level of sophistication. For the average chemist as well as the average student of chemistry, the simpler the explanation the better, as long as it has practical value and does not violate fundamental principles. Insistence on "rigor" at the expense of understanding has always seemed to me foolish and unnecessary. I truly believe that the models under discussion offer unique aid to understanding **not available with any other models or by any other means,** and that they are **fundamentally sound.**

Atomic Models

The models of atoms (B2, B21) are constructed of styrofoam spheres in a correct relative scale of nonpolar covalent radii. The exact dimensions will depend on the anticipated size of the audience. They are colored to represent electronegativity, red meaning very low, yellow intermediate, and blue very high. Thus the range in hues from red through red-orange, orange, yellow-orange, yellow, yellow-green, green, blue-green, and blue represents the change from very low to very high electronegativity. Attached to the surface in appropriate positions are much smaller styrofoam spheres representing the outermost electronic structure. These are in pairs representing the outer orbitals, and colored black to represent a vacancy and white to represent an electron.

A pair of black spheres corresponds to a vacant orbital, a pair of white spheres a lone pair of electrons, and a pair of white and black correspond to a bonding orbital capable of forming one covalent bond. The electrons are distributed as they would be for bonding, (representing the "valence state") and the orbitals are directed as expected from repulsions among electron pairs. For example, a model of an M2 or M2' element has two outermost electrons, so these are paired with black vacancies at opposite sides of the atomic sphere. For boron the three bonding orbitals are directed toward the corners of an equilateral triangle with the nucleus at its center, and the fourth orbital, which is vacant, is placed on one side of this triangle, as until occupied it can have no effect on the bond angles.

From M4 on, the atomic models have the four filled or half-filled orbitals at the corners of a regular tetrahedron.

A set of such atomic models provides a most vivid illustration of the periodic law (B36). From the models spread out at random they are picked out one by one in order of increasing atomic number. Wire hooks permit linking the atoms together in a chain. Models of the atoms of the first twenty elements will suffice for this demonstration. When they are aligned from 1 to 20, students can at once see that a periodicity exists. Following hydrogen and helium, lithium is represented by a relatively large red sphere, beryllium smaller and orange, boron smaller and yellow-orange, carbon smaller and greenish yellow, nitrogen very slightly smaller and green, oxygen very slightly smaller and blue-green, and fluorine slightly smaller and blue. Then comes neon, very slightly smaller and black (the color selected for the M8 elements which have no significant electronegativity). It is followed by a trend from sodium to argon very similar to that of Li-F, followed again by a large red sphere representing potassium, and a somewhat more orange-red sphere representing calcium. The number of outermost electrons can easily be seen to change from 1 to 2, hydrogen to helium, and then 1 to 8 and 1 to 8 in the second and third period elements.

Students can then be shown that if the two periods of 8 are disconnected from one another and placed parallel so that corresponding parts of each period are adjacent, not only are the **trends** from one extreme to the other clearly depicted, but also the **beginnings of each group of similar elements** are visible. A complete set of such models hung from a framework on which their symbols and atomic numbers are painted makes a remarkably useful three-dimensional periodic table--the atomic models being three-dimensional, the table planar.

These models can then be used individually in discussing all kinds of major group chemistry. First, the relationship of atomic structure to the physical state of the element can be discussed. Then predictions can be demonstrated for any combination of metal with nonmetal. Such predictions can then be verified by demonstration of molecular models as next described.

Molecular Models

Similar styrofoam models of molecules can be made easily and cheaply (B2, B20, B37, B38). Here it is desirable to estimate the radii of the combined atoms according to their partial charges, such that the sum will provide the correct bond length, to scale. There would be no point in representing electronegativity in these models because the individual atoms within a molecule are equal in that respect. What is important is to represent the partial charges which have resulted from the equalization process. This can be done by an entirely different color scheme if desired. It seems satisfactory to me to relate the electronegativity to partial charge--that is, to give to an atom with high partial negative charge approximately the same hue as indicated its initially high electronegativity in the atomic model. Fluorine atom is blue, and highly negative fluorine in fluorides is blue or blue-green. Sodium atomic model is red, indicating low electronegativity, and sodium with high partial positive charge is also red. Thus the range of increasing partial negative charge is represented as a range from yellow through yellow-green and green and blue-green to blue. The range of increasing partial positive charge is represented as a range from yellow through yellow-orange, orange, red-orange, and red.

Such models retain the unused lone pairs and vacant orbitals on the appropriate atoms. They show single bonds by tangential contact, and multiple bonds by cutting off segments of each sphere to produce the proper bond length when the flat sections are then glued together. The atoms are of course glued together at the proper bond angles.

The atomic models can be used to make predictions and the molecular models to verify them. In this way, students can learn that **there exists a direct, logical, understandable relationship between the nature of the individual atoms and the properties of the molecules of which they are components.** This beautiful cause-and-effect relationship is **at the very heart of chemical science,** and **its recognition provides the sorely needed bridge between theory and practice.**

For example, a model of an atom of hydrogen shows by the greenish-yellow color that the electronegativity is just slightly above the median value, and by the half-filled orbital that it can form one covalent bond. The model of the oxygen atom reveals its high electronegativity by the blue-green color and the outermost electron level is shown to consist of two lone pairs of electrons and two half-filled orbitals. Because these are distributed roughly at the corners of a tetrahedron, the bonding orbitals are not opposite one another but form an angle of about 109°. It is very easy to predict from these atomic models that one oxygen atom can form covalent bonds to two hydrogen atoms, completely filling all the normal covalent bonding potential of each. The two O-H bonds should be at an angle near to 109°, except that lone pair electrons clearly visible on the oxygen model should exert somewhat greater repulsions than shared pairs. This would make the angle a little smaller than 109°. Since the oxygen is obviously more electronegative than the hydrogen, the bonds are predicted to be polar covalent, with oxygen partially negative and hydrogen partially positive. With positive hydrogen and with negative oxygen having lone pair electrons, formation of protonic bridging between molecules can also be predicted. From the presence of two lone pairs of electrons and two positive hydrogen atoms on each oxygen, one can predict that the maximum number of protonic bridges formed by each oxygen can be four, and when these are all formed at once, a solid structure should result which is not very closely packed since normally, particles can accommodate more than four similar particles around them. Preparation is thus made for explaining why ice is less dense than liquid water at the same temperature, wherein the protonic bridging is somewhat broken down, permitting closer packing.

Exhibition of a model of a molecule of water will then show how accurate the predictions from the atomic models have been. It will show the hydrogen atoms as orange, having partial positive charge, and the oxygen atom to be green, having partial negative charge. The bond angle will be predicted, and the lone pairs on the oxygen will be evident. Two such models can be used to indicate protonic bridging and show the requirements for such bridging. A model of ice will confirm the prediction made for the structure. The chemical properties of water can also be discussed from the model. The electron pairs on the negative oxygen will make the molecule an electron pair donor, and the positive hydrogen will allow it to undergo protonic reactions. If the calculation of bond energy is added to this demonstration, an unparalleled understanding of water is made available.

E X P L A I N I N G   C H E M I S T R Y :
GENERAL PROCEDURE

1) SHOW **ATOMIC** MODELS OF APPROPRIATE ELEMENTS,
   Pointing out features of **size, electronegativity,** and **outer electronic arrangement,** and their influence on the **bonds** such atoms might be expected to form.

2) SHOW MODELS OF THESE **ELEMENTS** IN THEIR STAND-
   ARD STATES, and explain how these states are an inevitable (or at least reasonable) result of the nature of the INDIVIDUAL ATOMS. If reasonable, also show samples of these elements, pointing out that the model represents a greatly enlarged view of the actual element. State **how much energy** is needed to **atomize** the element as a prerequisitve to the **rearrangement** of its atoms.

3) SHOW MODELS OF EACH OF THE **REACTANT**
   SUBSTANCES and show how their **formulas and structure** result from the nature of their **individual atoms.** If possible, also show actual samples of these substances, to tie these models to reality. Explain the physical state of these substances. CALCULATE THE BOND AND ATOMIZATION ENERGIES of each reactant, using the **theory of polar covalence.**

4) BASED ON THE NATURE OF THE REACTANTS, make
   predictions as to possible favorable rearrangements of their atoms to form new, more stable substances.

5) SHOW MODELS OF EACH OF THE REACTION PRODUCTS
   and explain them as in (3). If possible, show **actual samples** of each product. CALCULATE THE BOND AND ATOMIZATION ENERGIES of each product.

6) WRITE THE CHEMICAL EQUATION and beneath it, the
   ATOMIZATION ENERGIES of reactants and products.
   DERIVE the **heat of reaction.** Make rough predictions about probable entropy changes. Then explain **WHY** this rearrangement of atoms occurs.

7) If possible, then DEMONSTRATE THE ACTUAL REACTION.
   Optionally, this might well be step (1).

It is difficult to imagine any aspect of major group chemistry which could not be explained more lucidly with the help of such atomic and molecular models. Another particularly good example is provided by a set of molecular models of binary compounds of hydrogen across the periodic table. Students seeing the transition from large blue-green hydrogen to small red-orange hydrogen can understand the remarkable transition in properties much more clearly, when they can tell by looking at the models which are hydrides and which are compounds of neutral or positive hydrogen. They can easily and confidently predict the physical and chemical properties of these compounds. In general the **periodicity of the chemistry** of the major group elements is very vividly displayed when the molecular models are appropriately arranged.

Such models have the commercial disadvantage that atoms of a particular element are not interchangeable among several combinations. A separate model is needed for each compound. However, the models are very inexpensive and students can learn much by building a set for future use by the instructor. In my own experience, student reaction has been very favorable both toward the models and toward the general approach to explaining chemistry. Many senior chemistry majors have volunteered that although they had been required to commit much of chemistry to memory, this particular approach was the first in their chemistry studies which had really tried to explain chemistry to them. If this present discussion could stimulate a renaissance in the use of these models, I am convinced that students would be most appreciative and that chemistry teachers would find their task much easier and more satisfying. If use of such models could lead to any objectionable results I am not aware of it. The only risk, which seems extremely slight, is that students whose completion of high school was more purge than passing might think that sodium is red and fluorine blue because their atomic models are so painted.

This is probably my final appeal for rational and fair minded consideration of such models. Since their potential first became evident, I have done all within my limited physical ability to encourage their use, by both writings and lecture demonstrations. Lacking the energy and time to satisfy all the early demands for personal demonstrations, I made, with aid from the National Science Foundation, three filmed lecture demonstrations of the models (B37-39) which have been widely viewed around the world. Several more films were planned and aid was promised, but support was withdrawn because of politics. Nevertheless the early films do suggest the possibilities, which are limited only by the ingenuity and imagination of the instructor. Now that the validity of the partial charges has become so firmly established, it is difficult to see any logical basis for withholding this valuable aid from students.

Recurrently there are worries and accusations about the quality of chemistry teaching and the failure of chemistry to attract some of the brightest minds. Worldwide conferences become more frequent, but to what purpose, if they cannot overcome the resistance to change? We are not even teaching as well as we now know how to do. Who knows how much more attractive chemistry as a profession might become if it were presented in a more logical and meaningful manner?

My final (?) plea to all teachers of chemistry is to try to break loose from tradition and expose your students to opportunities of which they are now deprived. Refuse to accept what has always been done as a standard of performance, and I beg of you, **don't** teach the way you were taught. Accept improvement, but refuse to admire the Emperor's new clothes unless you yourself can actually see them clearly. Try to put aside what everyone does or what everyone thinks and make a few independent comparisons, using your own good judgment.

Compare, for example, any molecular orbital diagram or electron density contour drawing with the appropriate atomic and molecular models just described, and decide which really provides better understanding to the average student, or the average chemist. Compare the customary practice of presenting the periodic law and table as a set of rules, with the much deeper understanding so easily potentially derivable from an inexpensive set of atomic models. Compare the plain, uninspiring recital of descriptive chemistry with the simple exposition of the relationship between the qualities of atoms and the properties of their compounds. Compare the presentation of a chemical reaction as a stark equation to be learned by rote, with a simple analysis of the bond energies involved and a quantitative understanding of the origin of these energies, so that the rearrangement of atoms that is represented by the equation has purpose and meaning.

All my career I have respected and admired the theoretical chemists for their extraordinary intellectual and mathematical talents, and humbly wished I could really understand their methods and procedures. But I have also been alert to the results of these procedures. When it takes many pages of results of very complex calculations to describe the very simplest of chemical bonds in the H_2^+ ion, and when thousands of dollars are spent for computer time to obtain results which may, if a large number of arbitrary parameters were luckily chosen, approach accuracy within fifty per cent, I can appreciate that all this may be necessary for the eventual advancement of knowledge, but I am unable to find much that is applicable to practical understanding by ordinary chemists and students.

At the very beginning of my teaching career I found chemistry very frustrating to teach because there were so few bridges between theory and practice. As a teacher I felt impelled to explain chemistry, and explanations were not available. In fact, most of the seemingly obvious questions appeared not even to have been asked. Partly in desperation at my own limitations, and partly in recognition of the practical limitations of quantum mechanics, I embarked on a series of studies designed to make chemistry simple enough so I myself could understand it. I think I have been blessed with a significant degree of success. I have the illusion of understanding chemistry far better than I ever thought possible.

Now my major frustration lies in my difficulty in persuading my fellow chemists that they, too, and all their students, can profit from this approach. I am sure they could if they would only break loose from tradition and make the great effort needed to regain the viewpoint of their students. Chemistry is neither a miscellaneous collection of enormous numbers of facts with occasional guesses about their meaning, nor a solemn exploration of mathematical complexities defying simple solution and revealing secrets only to the chosen few. Chemistry is actually a beautiful science, displaying in a simple and understandable way a perfectly logical and delightfully consistent cause-and-effect relationship between the structure of atoms and the properties of the chemical substances which they form. I hope this book will help all chemists to appreciate and to realize the possibilities for greater understanding.

A. GENERAL REFERENCES

1) J. C. Slater, Phys. Rev., **36**, 57 (1930)
2) E. Clementi and D. L. Raimondi, J. Chem. Phys., **38**, 2686 (1963)
3) H. Allred and E. G. Rochow, J. Inorg. Nucl. Chem., **5**, 264, 269 (1958)
4) CODATA, J. Chem. Thermodyn., **8**, 603 (1976); CODATA Bulletin, 1978
5) National Bureau of Standards Technical Notes 270-8
6) L. Brewer and G. M. Rosenblatt, Adv. High Temp. Chem., **2**, 1 (1969)
7) L. Pauling, "Nature of the Chemical Bond ," Cornell Univ. Press, Ithaca, N.Y., 2nd Ed. 1940; 3rd Ed. 1960
8) R. G. Parr, R. A. Donnelly, M. Levy, and W. E. Palke, J. Chem. Phys., **68**, 3801 (1978)
9) P. Politzer and H. Weinstein, J. Chem. Phys., **71**, 4218 (1979).
10) K. S. Pitzer, J. Am. Chem. Soc., **70**, 2140 (1947)
11) R. S. Mulliken, J. Am. Chem. Soc., **72**, 4493 (1950); ibid **77**, 884 (1955)
12) JANAF Thermochemical Tables, Dow Chemical Co., Midland, Michigan.
13) J. D. Cox and G. Pilcher, "Thermochemistry of Organic and Organometallic Compounds," Academic Press, N. Y., 1970.
14) D. R. Stull, E. F. Westrum Jr, and G. C. Sinke, "The Chemical Thermodynamics of Organic Compounds," John Wiley, New York, 1969
15) J. A. Kerr and A. F. Trotman-Dickenson, "Strengths of Chemical Bonds," in "Handbook of Chemistry and Physics," 61st Ed., Chemical Rubber Co., Cleveland, 1980-1, p F 222 et seq.
16) M. Atri, R. R. Baldwin, G. A. Evans, and R. W. Walker, J. C. S. Faraday Trans. II, **74**, 366 (1978)
17) C. E. Canosa and R. M. Marshall, Int. J. Chem. Kinetics, **13**, 303 (1981)
18) T. N. Bell, K. A. Perkins, and P. G. Perkins, J. Phys. Chem., **83**, 2321 (1979)
19) B. A. Robaugh and G. D. Stein, Int. J. Chem. Kinetics, **13**, 445 (1981)
20) S. W. Benson, P. A. Knoot, and S. P. Heneghan, Int. J. Chem. Kinetics, **13**, 518 (1981)
21) D. R. Lide, J. Chem. Phys., **33**, 1514, 1519 (1960)

22) C. T. Zahn, J. Chem. Phys., **2**, 671 (1934)

23) J. R. Platt, J. Chem. Phys., **15**, 419 (1947)

24) J. R. Platt, J. Phys. Chem., **56**, 328 (1952)

25) R. D. Brown, J. Chem. Soc., 2615 (**1953**)

26) M. J. S. Dewar and R. Pettit, J. Chem. Soc., 1625 (**1954**)

27) K. J. Laidler, Can. J. Chem., **34**, 636 (1956)

28) J. B. Greenshields and F. D. Rossini, J. Phys. Chem., **62**, 271 (1958)

29) T. L. Allen, J. Chem. Phys., **31**, 1039 (1959)

30) E. G. Lovering and K. J. Laidler, Can. J. Chem., **38**, 2367 (1960)

31) V. M. Tatevskii, V. A. Benerski, and S. S. Yarovoi, "Rules and Methods for Calculating the Physico-Chemical Properties of Paraffinic Hydrocarbons)" (translation B. P. Mullins, Ed.), Pergamon Press, Oxford, 1961

32) H. A. Skinner and G. Pilcher, Quart. Rev., **17**, 264 (1963)

33) J. W. Anderson, G. H. Beyer, and K. M. Watson, Nat. Petrol. News, **36**, R476 (1944)

34) J. L. Franklin, Ind. Eng. Chem., **41**, 1070 (1949)

35) M. Souders, C. S. Mathews, and C. O. Hurd, Ind. Eng. Chem., **41**, 1408 (1949)

36) S. W. Benson and J. H. Buss, J. Chem. Phys., **29**, 546 (1948)

38) P. Politzer, J. Am. Chem. Soc., **91**, 6235 (1969)

39) P. Politzer, Inorg. Chem., **16**, 3350 (1977)

40) "Interatomic Distances," Special Publication No. 11, The Chemical Society, London, 1958.

41) "Interatomic Distances Supplement," Special Publication No. 18, The Chemical Society, London, 1965.

42) W. V. Steele, Annu. Rep. Prog. Chem. Sect. A: Physical Inorg. Chem., **1974**, (pub. 1975), **71**, 103-118.

43) H. A. Skinner, in "Advances in Organometallic Chemistry," Academic Press, London, 1964, Vol. **2**, p 49.

B.   REFERENCES TO THE AUTHOR'S PREVIOUS WORK
(Having Special Relevance to This Book)
BOOKS

1) "Chemical Periodicity," Reinhold Publishing Corp., New York, 1960
2) "Teaching Chemistry with Models," Van Nostrand Co, Princeton, N. J., 1962
3) "Inorganic Chemistry," Reinhold Publishing Corp., New York, 1967
4) "Chemical Bonds and Bond Energy,", Academic Press, New York, 1st Ed. 1971; 2nd Ed. 1976
5) "Chemical Bonds in Organic Compounds," private edition, 600 copies distributed to libraries and organic chemists at major institutions, 1976
6) "Explaining Chemistry Better," private edition 1000 copies distributed to chemistry faculty at more than 450 colleges and universities, on request; 1980.

ENCYCLOPEDIA ARTICLES

"Encyclopedia of Chemistry," 3rd Ed., C. A. Hampel and G. G. Hawley, Eds., Van Nostrand Reinhold, New York, 1973; the following articles:
7) "Bonding, Chemical," p 152-8
8) "Electronegativity," p 385-6
9) "Halogen Chemistry," p 520-1
10) "Hydrogen Chemistry," p 546-8
11) "Oxide Chemistry," p 778-781

JOURNAL ARTICLES

12) "An Interpretation of Bond Lengths and a Classification of Bonds," Science, **114,** 670-2 (1951)
13) "Electronegativities in Inorganic Chemistry," J. Chem. Educ., **29,** 539-44 (1952)
14) same, II, **31,** 2-7 (1954)
15) same, III, 31, 238-45 (1954)
16) "Partial Charges on Atoms in Organic Compounds," Science, **121,** 207-8 (1955)
17) "Electronegativities in Inorganic Chemistry. A Revision of Atomic Charge Data," J. Chem. Educ., **32,** 140-1 (1955)
18) "A Comparative Study of Methyl Compounds of the Elements," J. Am. Chem. Soc., **77,** 4531-2 (1955)

19) "New Molecular Models Showing Charge Distribution and Bond Polarity," J. Chem. Educ., **34,** 195 (1957)

20) "Models for Demonstrating Electronegativity and Partial Charge," J. Chem. Educ., **36,** 507–512 (1959)

21) "Atomic Models in Teaching Chemistry," J. Chem. Educ., **37,** 307–310 (1960)

22) "Principles of Chemical Bonding," J. Chem. Educ., **38,** 382–91 (1961)

23) "Principles of Chemical Reaction," J. Chem. Educ., **41,** 13–22 (1964)

24) "Principles of Hydrogen Chemistry," J. Chem. Educ., **41,** 331–3 (1964)

25) "Principles of Halogen Chemistry," J. Chem. Educ., **41,** 361–6 (1964)

26) "Principles of Oxide Chemistry," J. Chem. Educ., **41,** 415–20 (1964)

27) "Bond Energies," J. Inorg. Nucl. Chem., **28,** 1553–1565 (1966)

28) "The Role of Coordinate Covalence in Nonmolecular Solids," in Advances in Chemistry Series, No. 62, R. F. Gould, Ed., Am. Chem. Soc., Washington, D. C., 1967; p 187–202

29) "The Nature of 'Ionic' Solids: The Coordinated Polymeric Model," J. Chem. Educ., **44,** 516–523 (1967)

30) "Multiple and Single Bond Energies in Inorganic Molecules," J. Inorg. Nucl. Chem., **30,** 375–93 (1968)

31) "Why Does Methane Burn?" J. Chem. Educ., **45,** 423–5 (1968)

32) "Recent Improvements in Explaining the Periodicity of Oxygen Chemistry," J. Chem. Educ., **46,** 635–9 (1969)

33) "Chemical Bonds. I" J. Coll. Sci. Teaching, **1,** 16–23 (1972)

34) "Chemical Bonds. II" J. Coll. Sci. Teaching, **1,** 47–52 (1972)

35) "Radical Reorganization and Bond Energies in Organic Molecules," J. Org. Chem., **47,** 3835–3839 (1982)

36) "On the Significance of Electrode Potentials," J. Chem. Educ., **43,** 584–6 (1966)

FILMS

45 min, 16 mm sound-color, Extension Div., Univ. Iowa, 1959

37) "Atomic Models, Valence, and the Periodic Table"
38) "New Models of Molecules, Ions, and Crystals, Their Construction and General Use in Teaching Chemistry"
39) "A Special Set of Models for Introducing Chemistry"

C. REFERENCES TO CHEMICAL BONDS AND BOND ENERGY
(from Science Citation Index)

1971
1) Bauman, A. J., Separat. Sci., **6,** 715
1972
2) Hojer, G., Act. Chem. S. C., **26,** 3723
3) Shaw, R. W., Phys. Rev. B, **5,** 4856
1973
4) Jolly, W. L., J. Am. Chem. Soc., **95,** 5442
5) Kutoglu, A., Act. Cryst. B, **B29,** 2891
6) Lewis, J. E., Phys. St. S.A. 16,161
7) Mittal, L. J., J. Phys. Chem., **77,** 1482
8) Seals, R. D., Inorg. Chem. N., **12,** 2485
9) Parry, D. E., Chem. P. Letters, **20,** 124
10) Mikler, J., Monats. Chem., **104,** 376
11) Phillips, J. C., J. Phys. Ch. S., **34,** 1051
12) Johnson, O., Inorg. Chem., **12,** 780
1974
13) Wright, J. S., Theor. Chim., **36,** 37
14) Carver, J. C., J. Am. Chem. Soc., **96,** 6851
15) Kudo, H., B. Chem. S. J., **47,** 2162
16) Wright, J. S., J. Am. Chem. Soc., **96,**4753
17) Smith, V. H., Chem. Phys., 5, 234
18) Pradel, P., Nucl. Instr., **121,** 111
1975
19) Joshi, R. M., J. Macr. S. Ch. R. A. **9,** 1309
20) Vasini, E., Z. Phys. Ch. F., **94,** 39
21) Madhukar, A., Sol. St. Comm., **16,** 383
22) Maekawa, T., J. Chem. Phys., **62,** 2155
23) Allen, L. C., J. Am. Chem. Soc., R, **97,** 6921
24) Dickenson, T., J. Solid St. Ch., **13,** 237
25) Elson, T. H., J. Am. Chem. Soc., **97,** 335
26) Hess, P. C., Geoch. Cor. A, **39,** 671
27) Kulkarni, K. S., I. J. PA phys. L, **13,** 780
28) Ludska, R., Phys. Rev. L., **34,** 1170
29) Seno, M., B. Chem. S. J., **48,** 2001
30) Gombler, W., Z. Naturfo B, **B30,** 169
31) Szabo, Z. G., Act. Chim. H., **86,** 127
1976
32) Jarvis, B. B., J. Org. Chem., **41,** 1557
33) Haneman, D., ACS Sym. S. **1976,** 157
34) Ludska, R., Phys. Rev. B., **13,** 739

35) Maloney, K. M., J. Inorg. Nucl. Chem., **38,** 49
36) Sazanov, Y. N., Thermoch. Act., **15,** 43
37) West, R., J. Am. Chem. Soc., **98,** 5620
38) Zebrowski, J., Act. Phys. P.A., **50,** 307
39) Appelbau, J. A., Phys. Rev. B, **14,** 588
40) Appelbau, J. A., Phys. Rev. I, **36,** 168
41) Gombler, W., Z. Naturfo B, **31,** 727
1977
42) Hahn, S. K., High Temp. S, **9,** 165
43) Thakur, L., I. J. PA Phys. W, **15,** 732
44) same, 794
45) Macios, A., An. Quimica, **73,** 412
46) Gray, R. C., J. Elec. Spec., **12,** 37
47) Maslingi, I., Czech J. Phys., **27,** 1389
48) Hilpert, K, Ber. Bun. Ges., **81,** 30
49) Mague, J. T., Inorg. Chem., **16,** 131
50) Yuankai, W., Geochimica **1977,** 65
51) Sinha, A. K., I. J. PA Phys, N **15,** 115
52) Jeffries, J. A., J. Chem. S. DA, **1977,** 1624
53) Allen, L. C., Adv. Mol. Int., **11,** 233
54) Davidson, R. B., J. Chem. Educ., **54,** 531
55) Fisher, G. B., Surf. Sci., <u>65</u>, 210
1978
56) Detar, D. F., J. Am. Chem. Soc., **100,** 2484
57) Hayes, D. N., J. Am. Chem. Soc., **100,** 4378
58) Benson, S. W., Angew. Chem., **17,** 812
59) Evans, D. K., Chem. Phys., **32,** 81
60) Leeuw, D. M. D., Chem. Phys., **34,** 287
61) Porte, L., Chem. P. Lett., **56,** 466
62) Sarade, P. R., Z. Naturfor A, **33,** 946
63) Maekawa, T., B Chem. S. J., **51,** 780
64) Lemke, B. P., Phys. Rev. B, **17,** 1893
65) O'Keeffe, M., Acta Cryst. B, **34,** 27
66) Reffy, J., Act. Chim. H., **96,** 95
67) Woolfe, A. A., J. Fluorine, **11,** 307
68) Misawa, M., J. Phys. Jpn, **44,** 1612
69) Schultke, H., Z. Anorg. A. C. **445,** 20
70) Gombler, W., Z. Anorg. A. C. **439,** 193
71) Thakur, L., I. J. Phys. A, **52,** 521
72) Garrett, M. K., Nature, **274,** 913
73) Jacobs, P. A., J. Inorg. Nucl. Chem., **40,** 1919
74) Hellwink, D., Chem. Zeitung, **102,** 369
75) Montgomery, R. L., J. Chem. Ther. **10,** 471

76) Mortier, W. J., J. Catalysis, **55**, 138
77) Parr, R. G., J. Chem. Phys., **68**, 3801
78) Bowen, R. L., J. Dent. Res., **57**, 255
79) Mezey, P. G., Chem. Phett., **59**, 117
1979
80) DeLeeuw, D. M., Chem. Phys., **38**, 21
81) McAlpine, R. D., Chem. Phys., **39**, 263
82) DeLeeuw, D. M., Chem. P. Lett., **61**, 191
83) Ray, N. K., J. Chem. Phys., **70**, 3680
84) Juhlke, T. J., J. Am. Chem. Soc., **101**, 3229
85) Liu, K., J. Phys. Chem., **83**, 970
86) Politzer, P., J. Chem. Phys., **71**, 4218
87) Barone, V., J. Chem. S. P2, **1979**, 1309
88) Joshi, R. M., J. Macr. S. Ch., **A13**, 1015
89) Chien, J. C. W., Eur. Polym. J., **15**, 1059
90) Lercher, J., Z. Phys.Ch. W, **118**, 209
91) Vinek, H., Z. Phys. Ch. W, **118**, 192
92) Schnering, H. G., Z. Anorg. A.C., **456**, 194
93) Misicyu, M., J. Chem. S. P2, **1979**, 1160
94) Yang, R. T., Alche J., **25**, 811
1980
95) Elkaim, J. C., J. Phys. Chem., **84**, 354
96) Eades, R. A., J. Chem. Phys., **72**, 3309
97) Greenspan, A. D., Int. J. Gen. S., **6**, 25
98) Vinek, H., Z. Phys. Ch. W, **120**, 119
99) Ishiwata, M., J. An. Ap. Pyr., **2**, 153
100) Lagow, R. J. book 14202, R23 177
101) Appelbau, J. A., book 12446 R, **19**, 43
102) Singh, P.P., J. Inorg. Nucl. Chem., **42**, 521
103) Giridhar, A., J. Non-Cryst., **37**, 165
104) Drache, M., B. S. Ch. Fr. 1, **1980**, 105
105) Ohwada, K., J. Chem. Phys., **72**, 3663
106) Stewart, R.F., Am. Mineral, **65**, 324
107) Pytela, O., Coll. Czech, **45**, 1269
108) Yamamoto, M., Kobunsh Ron, **37**, 319
109) Vinek, H., React. Kin. C, **14**, 273
110) ? R. A., J. Chem. Phys., **72**, 3309
111) ? K. W., J. Phys. Chem., **84**, 3172
112) March, N. H., P. Nas. Phys., **77**, 6285
113) Bell, N. A., Act. Cryst. B, **36**, 2950
114) House, J. E., Inorg. Nucl. **16**, 543
115) Gerasimo, S. B., J. Appl. Chem., **53**, 870
116) Dudnikov, T. A., J. Struct. Ch., **21**,724

1981
117) Brant, P., J. Am. Chem. Soc., **103,** 329
118) Sabbah, R., Thermoc Act, **43,** 269
119) Von Schne, H. G., Angew. Chem, **20,** 33
120) Briggs, A. G., J. Chem. R-S, **1981,** 69
121) Wang, W. I., Appl. Phys. L., **38,** 708
122) Medepall, K. S., Chem. Eng. CO, **8,** 269
123) Inoue, Y., J. Org. Chem., **46,** 2267
124) Kingsbury, C. A., Phosphor SU, **9,** 315
125) Susu, A. A., J. Am. Oil Ch., **58,** 657
126) Trinquie, G., Chem. P. Lett., **80,** 552
127) Briggs, A. G., Spect. Act. A N, **37,** 457
128) Kier, L. B., J. Pharm. Sci., **70,** 583
129) Otsuka, K., J. Chem. S F 1, **77,** 2717
130) Lam, Y. W., J. Org. Chem., **46,** 4462
131) Mohan, R., I. J. Chem. A, **20,** 864
132) Green, D. E., P NAS Biol., **78,** 5344
133) Ohwada, K, Spect. Act. A, **37,** 381
134) Aribike, D. S., Thermoc Act., **47,** 1
135) March, N. H. Bk 20372 **4,** 92
136) Su, C. Y., J. Vac. Sci. T., **19,** 481
137) Rao, K. J., Sol. St. Comm., **39,** 1065
138) Bhandia, A. S., Surf. Sci., **108,** 587
139) Lin, M. C., Comb. Flame, **42,** 139
140) Robertson, J., Phil. Mag. B, **44,** 239
141) Woolf, A. A., Book 20404 R 24 1
142) Tatsumi, T., J. Orgmet. Ch. **218,** 177
143) Gruntz, K. J., Phys. Rev. B, **24,** 2069
144) Xu, G.X., Sci. Sinica, **24,** 956
145) Iwamoto, K., J. Non Cryst., **46,** 81
146) Dias, A. R., J. Org. Met. Ch., **222,** 69
147) Lindner, L., Radioch. Act., **29,** 9
148) Schwartz, J. A., Scrip Metal, **15,** 1309
149) Noller, H., Kem. Kozlem, **56,** 401
150) Lucovsky, G., J. Physique, **42,** 741
1982
151) Batsanov, S. S., Zh. Fiz. Khim., **56,** 320
152) Feltham, R. D., J. Am. Chem. Soc., **104,** 641
153) Labbe, P., Electr. Act., **27,** 257
154) Eltsov, A. V., Zh. Org. Kh., **18,** 168
155) Honevar, S., J. Catalysis, **73,** 205
156) Shaik, S. S., J. Am. Chem. Soc., **104,** 2708

(incomplete)

Index

(Note the following abbreviations: AE = atomization energy; BE = bond energy; EN = electro-negativity; PC = partial charge; f = figure; t = table.)